The Sixteenth Symposium of
The Society for the Study of
Development and Growth

Executive Committee, 1957

Developmental Cytology

JOHN R. PREER, JR. · CARL R. PARTANEN
T. C. HSU · GEORGE KLEIN AND EVA KLEIN
WOLFGANG BEERMANN · HANS F. STICH
DITER VON WETTSTEIN · DON W. FAWCETT
ALBERT L. LEHNINGER

Edited by
DOROTHEA RUDNICK

THE RONALD PRESS COMPANY · NEW YORK

Library of Congress Catalog Card Number: 55–10678

Foreword

*. . . jedenfalls müssen wir nach unseren Resultaten von einer phys-
iologischen Einheit der der Zelle zukommenden Chromosomen und
also von einer im "Kern" repräsentierten physiologischen Einheit
reden.*

THEODOR BOVERI

The Sixteenth Symposium of the Society for the Study of Development and Growth was organized around the cell itself. Advances
in biochemistry, immunochemistry, electron microscopy, as well as
in cytogenetics and cytochemistry, are rapidly altering our view of
cellular activity, as much by producing new solidly based factual
information as by changing the framework of permissible speculation regarding matters beyond our present means of observation.
From the many workers in these fields, it was possible to assemble
a series of speakers to discuss authoritatively new work on chromosomes, nucleoli, cytoplasmic organelles, cellular chemistry, and immune properties in relation to development or at least to cellular
differentiation. All this work involves techniques almost unknown
a few years ago. Since the Second World War these techniques have
been exploited, refined, and to an extent tamed. The present volume
containing the written versions of the Symposium papers constitutes
a report on the status of those aspects of cytology particularly interesting to students of development.

The Symposium was held at Kingston, Rhode Island, June 19–21,
1957. The Society is greatly indebted to the University of Rhode
Island, to the local committee and all others who entertained its
meeting so hospitably; to the National Science Foundation for a
grant supporting the Symposium, and to numerous colleagues who
have helped with editorial and other tasks.

DOROTHEA RUDNICK

Yale University
April, 1959

Contents

Developmental
Cytology

1

Nuclear and Cytoplasmic Differentiation in the Protozoa

JOHN R. PREER, JR.[1]

The protozoan geneticist finds useful the concept of the clone: the asexually produced descendants of a single protozoan. Although clones generally show great constancy, variations often arise. The problem of the origin and maintenance of such intraclonal variations is a long-standing one and has been the subject of many investigations. Since such studies encompass most of the questions raised by my topic, I shall limit my discussion to a consideration of some of the problems of intraclonal variation. It is significant (as many have pointed out) that the cells of a multicellular organism also constitute a clone and that the problem of cellular differentiation in multicellular organisms is therefore similar to the problem of variation within the clone in unicellular organisms.

Many variations are not inherited, being associated either with the different stages of the mitotic and meiotic cycles or arising from the action of differing environmental conditions. The changes induced by environmental differences, however, generally quickly disappear after a few generations when the variants are cultured under one set of conditions. This fact would indicate that the generation time is large in comparison to the time required for most biochemical reactions to reach their steady-state values so that new equilibria are rapidly attained. Furthermore, the loss or acquisition of substances accumulated or depleted as a result of differential environmental action should be rapid, for dividing cells undergo a complete molecular turnover in less than 60 generations. Thus the original molecules present in a cell used to start a clone will be progres-

[1] Zoological Laboratory, University of Pennsylvania. This work has been aided by grants from Phi Beta Psi Sorority and from the National Institutes of Health, Public Health Service.

sively diluted by cell division. After one fission the average number of the original molecules per cell will be reduced to $\frac{1}{2}$ (or 2^{-1}) times the starting number; after two fissions, to $\frac{1}{4}$ (or 2^{-2}), and after 60 fissions to 2^{-60} times the starting number. Since it can be simply shown [2] that the starting number is surely less than 2^{60}, the average number of original molecules per cell after 60 generations is less than one, and the probability of a given cell having an appreciable number of those molecules is consequently very low.

Many variations, however, persist for many cell generations under uniform environmental conditions, i.e., they are inherited. Other variants are not completely stable but undergo slow changes with time. The question can be argued as to whether variants of intermediate stability (the *Dauermodificationen* of Jollos or many of the changes associated with the meiotic cycle to be considered presently) should be classified as inherited or not. The label we attach to such phenomena, however, is clearly unimportant, while an understanding of the mechanisms of variation, whether of great or of intermediate stability, may be of considerable interest. We will consider a few of the major examples of stable and long-lasting changes.

Gene Mutations, Chromosomal Aberrations, and Changes in Ploidy

The best-known causes of stable variations in all organisms are gene mutations, chromosomal aberrations, and changes in ploidy.

Gene mutations in the protozoa, as in other organisms, are generally rare, haphazard, and nonadaptive. Thus they are not commonly encountered in clones of moderate numbers of cells. It might also be noted that only infrequently are gene mutations invoked as an explanation of cellular differentiation in multicellular organisms (see, however, Monod, 1947).

Chromosomal aberrations and aneuploid changes have been reported to occur with high frequency in the micronucleus of aging lines of *Paramecium* by Sonneborn and Schneller (1955a) and Dippell (1955). Such changes may also account for much of the nonviability occasionally encountered in clones of paramecia follow-

---

[2] A cubical cell 10^{-2} cm on edge with a specific gravity of 1.0 would weigh 10^{-6} gm and contain roughly 2^{60} molecules of molecular weight 1.0. Since the average molecular weight is clearly greater than 1.0, the actual number of molecules in the cell must be less than 2^{60}.

ing meiosis at autogamy and conjugation. Changes in ploidy fall outside the scope of the present discussion, since they are dealt with in other papers of this symposium. It is important to note that Sonneborn and Schneller (1955a) and Dippell (1955) found that in *Paramecium* such changes are deleterious and appear to be induced secondarily by other unknown hereditary changes associated with the ciliate life cycle.

The Ciliate Life Cycle

The ciliate life cycle has been the subject of investigation for many years by Maupas, Jennings, Woodruff, Sonneborn, and many others (see Jennings, 1944; Sonneborn, 1954). Typically, conjugation is followed by a period of immaturity during which animals cannot mate, then a period of maturity when animals are vigorous and mate readily, and finally a period of senescence during which the fission rate declines and conjugation or autogamy leads to nonviability.

Tens and even hundreds of cell generations may be involved in some cases. The very high degree of stability represented by these changes among the protozoa is better conceived by noting that the complex differentiation of an adult human from a zygote must require an average of only about 40 cell generations. (A single cell weighing 10^{-7} gm would, if all its progeny survived, in only 40 cell divisions produce 2^{40} cells weighing a total of 220 pounds.)

Three major kinds of changes occur during the meiotic cycle. First, in many ciliate stocks there is a sudden loss of the ability to mate following conjugation (but not following the almost identical uniparental process of autogamy!) and a return of the ability after many fissions. Second, in *Paramecium aurelia,* at conjugation and autogamy there is a suppression of the ability to undergo a new autogamy and a return of the ability after approximately 10 to 40 fissions. Third, in many ciliates the suppression of conjugation or autogamy leads, after many cell divisions, to reduced cell division and death, both before and after meiosis. Very little is known concerning the mechanisms of any of these phenomena. Sonneborn and Schneller (1955a) and Dippell (1955) have shown in *P. aurelia* that if aging has not proceeded too far, the postmeiotic nonviability in aged lines results from micronuclear chromosomal aberrations. The aberrations are not accumulated at a constant rate, however, but appear rapidly in old lines; this suggests that they are induced by the slow change of some inductive system which reaches a threshold. Sonneborn and Schneller (1955b) and Sonneborn *et al.* (1956) have presented evidence against the hypothesis that aging in *P. aurelia* results from the accumulation of chromosomal unbalance in the macronucleus; and Kimball and Gaither (1954) have excluded an accumulation of macronuclear lethals as the causative factor. The ultimate basis for life cycle changes in ciliates is unknown.

Mating Type Determination in Ciliates

At autogamy and conjugation in the ciliated protozoa the macronucleus starts degenerating and the one or more micronuclei undergo meiosis. Two haploid meiotic nuclei fuse to form a diploid syncaryon, which then undergoes a series of mitoses. Some of the mitotic products then differentiate into macronuclei, while others remain micronuclei. At subsequent cell divisions the macronuclei are segregated one to each animal, restoring the vegetative condition. Cytological observations on some ciliates reveal that the diploid number of chromosomes is increased many times in the formation of the macronucleus, and genetic evidence in others confirms this finding. Sonneborn (1937, 1954), working with *Paramecium*, and Nanney and Caughey (1953) with *Tetrahymena* have shown that mating type is determined by the macronucleus, while the macronucleus is itself determined at the time of its formation. Different macronuclear anlagen within the same reorganizing animal may be determined for different types. Environment and genes affect the frequency and range of types determined. In one group of varieties of *P. aurelia* the determination is strongly influenced by the old macronucleus through the cytoplasm so that the new type tends to be the same as the old. The situation resembles cytoplasmic inheritance but differs in that the cytoplasm appears to act only as a vehicle to transfer nuclear influences. The mechanism of the nuclear determinations is unknown, but Nanney (1956) has recently postulated alternative steady-state reactions (similar to those to be discussed presently in connection with antigenic inheritance in *Paramecium*) in the nucleus, sometimes spilling into the cytoplasm.

Self-reproducing Cytoplasmic Particles

Plastids. The inheritance of plastid variations in higher plant forms has been extensively studied. Many cases of variations in form and color of plastids have been shown to be under gene control; others are apparently due to changes in a cytoplasmic genetic system outside the plastids; and others are best interpreted as indicating that the plastid itself is a mutable genetic entity (cf. Rhoades, 1947, for a review). An interesting case in *Euglena* has been studied by Provasoli *et al.* (1951). They have shown that plastids (and in some cases the eyespot) may be destroyed irreversibly and with high frequency by streptomycin. Plastid loss may also be induced by temperature treatments (Pringsheim and Pringsheim, 1952). These findings may be regarded as evidence that the plastid is self-reproducing. On the other hand, it is also possible that the seat of the modified genetic system in *Euglena* is localized outside in the cytoplasm rather than within the plastid.

Kappa and Its Relatives. Kappa is a self-reproducing genetic entity found in the cytoplasm of *Paramecium* and is responsible for the liberation of a toxic substance. Kappa is gene-dependent (Sonneborn, 1943) but mutable (Dippell, 1950). Killers contain several hundred visible Feulgen positive particles, which have been shown to be identical with the genetic kappa particles (Preer, 1950). Mu (μ) particles (Siegel, 1953) are similar to kappa particles except that a toxic action is observed only at conjugation. Pi (π) particles (Hanson, 1954) are kappa mutants that show no toxic action. Other similar Feulgen positive particles (cf. Fauré-Fremiet, 1952) are found in other ciliates. While such bodies appear to be well integrated into the genetics and physiology of their bearers, they nevertheless appear to be highly specialized constituents; they are not general, even in the genus *Paramecium*.

Kinetosomes and Related Structures. Similar bodies play a role in mitosis and give rise to flagella and cilia. In some ciliates they also give rise to the primordia of the trichocysts. Kinetosomes are visibly self-reproducing. Visible self-reproduction, however, does not prove that a structure plays a genetic role in variation. It has been reported that in some forms the kinetosome arises *de novo* (Wilson, 1925, p. 388). However, *de novo* origin does not occur in others (Lwoff, 1950). Evidence against *de novo* origin of a kinetosomelike body is provided by the finding that chemical treatment of trypanosomes (Piekarski, 1949) induces irreversible loss of the parabasal body. This evidence is not decisive, however, for the site of the mutation may not be within the parabasal body but, instead, elsewhere in the cytoplasm or even in the nucleus.

It is of interest to inquire into the inheritance of variations in cilia and trichocysts. If they arise from mutable genetic elements, their mutation would give variants showing cytoplasmic inheritance. The specific mating-type substances of *Paramecium* are found on the cilia (cf. Metz, 1954). However, their variations are under nuclear control (Sonneborn, 1937, 1954a). The immobilization antigens are also found on the cilia in *Paramecium*. As we shall see presently, although both cytoplasm and genes are involved, the assumption of self-reproducing cytoplasmic particles does not help to explain the cytoplasmic and nuclear genetic system established here. We have recently found variant trichocysts within a subline of *P. aurelia*, variety 2, stock 197. The undischarged trichocysts are abnormal and variable in shape, unlike the regular carrotlike form of the original stock 197 or other stocks of paramecia (see Fig. 1). They also differ from the wild type in being very slow to discharge after animals are crushed beneath a cover-slip and observed with the phase microscope. The mutant was stable when selfed and gave a normal F_1 when crossed to normal stock 28 and other normal variety 2 stocks. A backcross to the mutant line gave 13 pairs with both exconjugant clones normal and 12 pairs with both mutant. An F_2 gave 37 pairs with both exconjugant clones normal and 7 pairs with both mutant. Thus both backcross and F_2 results are typical one-factor Mendelian ratios. A few pairs with one exconjugant clone normal and the other exconjugant mutant were found and were presumed to be due to cytogamy or macronuclear regeneration. The results, then, indicate that the abnormal trichocysts are due to mutation at a single locus. It is thus interesting to note that the two known cases of variation in the characteristics of cilia and the first case of trichocyst variation within a variety have provided no evidence for a genetic role of these structures or their primordia. More cases are needed.

Mitochondria. Cytological observations show that mitochondria often divide, but they also appear to fuse and disintegrate; there exist claims that they arise *de novo* as well. Their resemblance to plastid primordia has been cited as evidence that they, like plastids, may be self-reproducing. The best, but still only suggestive, evidence for their genetic role comes from the finding that cytochrome-deficient mutants in yeast and *Neurospora* are often cytoplasmically inherited (Ephrussi, 1953; Mitchell *et al.*, 1952, 1953). The cytochromes, of course, are known to be localized in the mitochondria of higher forms. No variant mitochondria have been reported in the protozoa; but genetic analysis of such variants, provided they can be found, would appear to be our best hope of ascertaining whether mitochondria have a genetic role.

STOCK 197

STOCK 28

Fig. 1. Sketches of the undischarged trichocysts from the aberrant stock 197 of *Paramecium aurelia* are shown in the upper row. Those of normal stock 28 are shown in the bottom row.

The Gullet. Conjugants of paramecia may be induced by antiserum to fuse, permanently but incompletely. Such doubles produce fairly stable clones of individuals with a double form containing two gullets, and one macronucleus (Sonneborn, 1950a; Margolin, 1954). These facts are most reasonably interpreted by assuming that the hereditary basis for the double condition lies within the cytoplasm and not in the nucleus. The question arises as to whether the whole cytoplasmic organization is responsible for the form or whether the mechanism can be reduced to one or more individual elements. Cytological observation might possibly suggest that the gullet is self-reproducing. Perhaps the double form is determined and maintained by the organization of the cytoplasm into the double form by the two self-reproducing gullets. Hanson (1955) has attempted to study the question by irradiating one of the two gullets in double animals with an ultraviolet microbeam. A double animal, with gullet damage so extensive that no visible irradiated gullet remains, may regenerate a new gullet, but more often it reorganizes as a single animal with one gullet—a process that sometimes occurs spontaneously. The interpretation is not clear, but the results show that a complete gullet is not necessary for a new gullet. The possibility that the gullet has a self-reproducing primordium, however, has not been ruled out.

Immobilization Antigens in *Paramecium*

Sonneborn (1947, 1948) showed that many stable serotypes can be isolated from a single clone of *P. aurelia*. Animals of a given serotype are immobilized when placed into an appropriate dilution of antiserum against that serotype but are unaffected when placed into antiserum against a different serotype. Thus, stock 51 yielded stable clones of serotypes A, B, C, D, E, G. Crosses between the serotypes within a stock revealed cytoplasmic inheritance. Stocks were found to vary in the spectrum of types they can manifest. Thus stock 29, which resembles 51 by having A, B, C, and D, did not manifest E or G but showed instead two other types, F and H. Crosses between stocks showed that the ability to manifest any given type, such as F, is due to a single Mendelian gene. Furthermore, in comparing different stocks, specific differences in serotypes, such as 29A and 51A, are often found. Such differences are also controlled by single alleles, one locus for each serotype. This picture, originally established for variety 4 of *P. aurelia,* has been confirmed and extended to several of the other varieties (cf. Beale, 1952; Finger, 1957a, 1957b; Pringle, 1956). Transformations from one serotype to another within a stock occur spontaneously but may also be induced by specific antiserum, radiation, changes in nutrition, temperature, and various chemical treatments.

The general situation may be summarized as follows: each organism contains a number of specificity determining loci, each concerned with one of the different serotypes. The specificity locus has, in the one case thus far investigated, been shown to be the same as the locus determining presence and absence (Reisner, 1955).

Usually one locus may come to expression at a time, i.e., only one serotype is manifested at a time. In most of the cases studied, any one of several different loci may come to expression under any one given set of environmental conditions, but once expressed, the continued expression of that locus, rather than other loci, is cytoplasmically inherited. The tendency to shift from the expression of one locus to a new one (transformation) is conditioned by this system of cytoplasmic inheritance, by the environment, the specificity locus and other loci (Beale, 1954).

Recently, studies on the immobilization antigens have been made by means of specific precipitation in gel (cf. Oudin, 1952). A technique of double diffusion in agar, studied quantitatively by Preer (1956), has been used by several workers. Antiserum is placed into the bottom of a small tube; agar is layered on top; and antigen is added on top of the agar. The antigens and antibodies diffuse into the agar, forming bands of precipitation where they meet, each separate antigen-antibody system generally forming a separate band. Relative concentrations of antigen and antibody may be found by taking advantage of the fact that the band position, p (the distance from antigen-agar interface to the band, divided by the total agar length), is linearly related to the logarithm of the ratio of the concentrations of antigen and antibody. By setting serial dilutions of antigen against constant antibody concentration, one obtains a linear relation that permits quantitative estimates of the antigen (see Fig. 2). Furthermore, information concerning diffusion

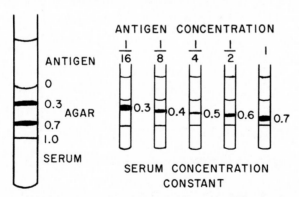

Fig. 2. A double diffusion tube is sketched on the left, showing band position, p, for bands at 0.3 and 0.7. The series of tubes sketched on the right shows the change in p obtained when serial dilutions of antigen are set against a constant antiserum concentration.

coefficients may also be obtained, for Ouchterlony (see Preer, 1956) showed that when the antigen and antibody are mixed in their immunological equivalence ratio, a stationary band is formed, and that

$$\frac{p}{1-p} = \sqrt{\frac{D_{ag}}{D_{ab}}}$$

where D_{ag} and D_{ab} are the diffusion coefficients of antigen and antibody respectively.

Finger (1956), using this technique, found that many antigens may be demonstrated in extracts of homogenates and of lyophilized animals of variety 2, serotype G. Extracts of serotype C had all these antigens but one, and in its place had another specifically different antigen. Thus each of the two serotypes was found to have a unique serotype-correlated antigen, for all the other antigens detected in animals of the two serotypes appeared identical. The almost perfect correlation between the antibody corresponding to these antigens and the presence of immobilizing activity of a large series of antisera against various serotypes leaves no doubt that he was working with the immobilization antigen or a closely associated substance.

Van Wagtendonk *et al.* (1956) report that if animals of variety 4 of *P. aurelia* are placed into 0.06M saline, a soluble antigen is liberated, which can be shown by the Oudin agar diffusion test to react with homologous and not heterologous antisera. Since the antigen studied by Finger was the *only* antigen detectable in his preparations that was correlated with serotype, it is likely that the antigen observed by van Wagtendonk *et al.* was the counterpart of Finger's antigen.

Our observations on variety 2 confirm the work of Finger. When an extract is diffused against an homologous serum, a series of bands are formed by the various antigens in the preparation. We designate them as antigen 1, antigen 2, antigen 3, etc. (see Preer and Preer, 1958, for a study of antigens 1–5). Only one of these antigens is found to be correlated with serotype: antigen 4. Antigen 4 is really a family of antigens. Thus serotypes C, E, and G have been found to contain a specific antigen 4 lacking in the others. We distinguish the different kinds of antigen 4 by subscripts, e.g., 4_C, 4_E, 4_G. The different antigen 4's are closely related, being present in approximately the same structures, in the same quantities, and having approximately the same diffusion coefficients. The antigens often cross-react weakly; and in one case, 4_G and 4_E, cross-react strongly.

The diffusion coefficient of antigen 4_G of variety 2 has been estimated at 2.4×10^{-7} cm²/sec (Preer and Preer, 1958). The antigen thus appears to be a very large molecule. It is inactivated by 80° C in 5 minutes. It is precipitated by cold TCA, by 80 per cent ethyl alcohol, and by ammonium sulfate at around one-half saturation at pH 7.0. It thus appears to have the properties of a protein. It is an excellent antibody producer, the best of any soluble antigen that we have been able to obtain from paramecia. We have studied its cellular localization (Preer and Preer, 1958). If paramecia are placed for 1 minute at about 15° C into a solution containing 15 per cent ethyl alcohol, 0.45 per cent NaCl, and 0.005M sodium phosphate buffer, pH 7.0, the cilia are shed and usually many trichocysts are discharged. (The same medium, but lacking alcohol, yields cilia and trichocysts, but not in such large numbers). The mixture is centrifuged for 1 minute at 1100G. The precipitate contains the deciliated intact animals overlain by a fluffy layer of trichocysts; the cilia may be recovered from the turbid supernatant by centrifugation for 2 minutes at 25,000G. Much antigen 4 can then be recovered from the cilia, little or none from the trichocysts, much in the deciliated bodies, and also much is found to have leaked into the salt alcohol medium (or salt medium, if the alcohol is omitted). Fractional centrifugation of homogenates reveals that much of the antigen is in the cilia and also that large quantities are associated with the body wall. These findings agree with the recent work of Beale and Kacser (1957) using fluorescent antibody: much of the immobilization antigen is on the cilia but even larger quantities are associated with body wall structures. The cilia extracts are remarkable in that we have detected only one antigen from them thus far: antigen 4.

Balbinder and Preer (unpublished), using agar diffusion, have studied the process of transformation of serotype. The variety 2 stock 28 serotype G is stable at 12° C and transforms to E when placed at 31° C. The transformation is complete in two days when the culture is fed in a manner to produce one fission daily, with all animals transforming synchronously, as judged by their uniform loss of ability to react with G antiserum and acquisition of ability to react with E antiserum during the transformation process. The two antigens 4_G and 4_E cross-react rather strongly, and it has thus far been impossible to study quantitatively the loss of the 4_G. The acquisition of 4_E, however, may be followed with the E serum P#132. This was prepared by injecting trichocysts and cilia into a rabbit. It was found to have titer against the heat stable trichocyst antigens 1 and 2 and the serotype correlated heat labile antigen 4. It was rendered specific for 4 by absorption with an extract of boiled

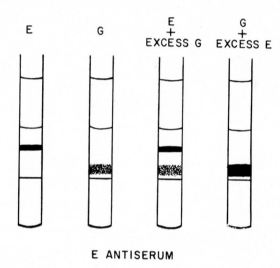

E ANTISERUM

Fig. 3. The reaction of E antiserum with antigens prepared from E, G, and mixtures of the two antigens. See text for details.

paramecia. The serum reacts strongly with 4_E, as shown in Fig. 3, and weakly with 4_G. Mixtures of the two antigens give two bands with the serum when the 4_G antigen is in excess; this indicates that the serum has some antibodies capable of reacting with 4_G and 4_E, and other antibody capable of reacting only with 4_E (cf. Preer, 1956, for a discussion of cross-reacting antigens in double diffusion reactions). When different dilutions of various ratios of the two antigens are run against serum P#132, it is found that p for the E band and p for the E + G band (only one band is obtained when mixtures containing an excess of E are set) gives an accurate logarithmic measure of the amount of 4_E in any mixture. Tests on transforming animals reveal two distinct bands and show no evidence for molecules of intermediate specificity during transformation. The results show a gradual loss of 4_G and synthesis of new 4_E. Fig. 4A shows the p values for 4_E in an experiment in which animals were shifted from 12° C to 31° C at zero time and fed for one fission daily with samples being removed periodically. The graph affords a logarithmic plot of the newly synthesized antigen. The line has been drawn by eye. Fig. 4B shows the amount of antigen in arbitrary units; the line is from 4A. Fig. 4C shows the lowest concentration of E serum immobilizing the animals during the transformation. It is noted that the kinetics of synthesis of new antigen follows a sigmoid

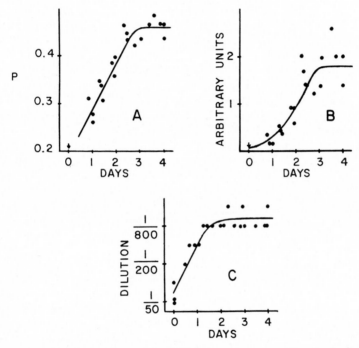

Fig. 4. Paramecia growing at 12° C were serotype G. At 0 time they were placed at 31° C, where they transformed to E. (A) shows E band position (p) during transformation. (B) shows the amount of E antigen computed from p values in (A). (C) shows the weakest concentration of E serum partially immobilizing animals during the transformation; it is noted that G animals cross-react with the E serum when it is in high concentration.

"growth" curve, similar to the kinetics of adaptive enzyme formation.

What is the mechanism of serotype determination in *Paramecium*? How can several serotypes be stable under the same conditions and also the role of the genes be satisfied? Mutable self-reproducing bodies have been ruled out (Sonneborn, 1950b) by the demonstration that the specific nature of the serotype being manifested in hybrids and during breeding tests always very precisely reflects the specific alleles in the nucleus. Other explanations have been reviewed by Beale (1954, 1957).

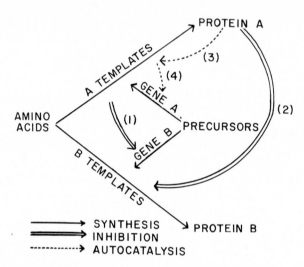

Fig. 5. Hypothetical steady-state mechanisms to explain the synthesis of proteins correlated with serotype in *Paramecium*. Four possibilities are indicated: (1) and (2) assume inhibitions and (3) and (4) assume autocatalysis with substrate competition.

Since the evidence presented here suggests that specific proteins are correlated with serotype, let us look at some of the possibilities in terms of the template hypothesis of protein synthesis. (For a discussion see Spiegelman, 1957.) Assume that each specific serotype locus, such as the genetic A locus, synthesizes A templates from appropriate precursors and that these make protein A directly from amino acids (see Fig. 5). B, C, etc., are synthesized in a similar fashion. There is presumably a common pool of precursors for the various templates as well as a common amino acid pool for protein synthesis. The observed cytoplasmic inheritance and sigmoid transformation curve can be explained by various hypotheses — four are shown, all of which may be considered alternative steady states in the sequence of reactions leading to A, B, C, etc. These steady states may operate by mutual inhibitions (cf. Delbruck, 1949) at any one of the various points. Thus there may be (1) an inhibition of heterologous genes by the templates, or (2) an inhibition of the heterologous templates by the final proteins. The steady states could also operate autocatalytically with (3) catalysis of

homologous templates by the final proteins (suggested by Spiegelman to explain adaptive enzyme formation in yeast), or (4) catalysis
of the activity of the gene by the homologous template (equivalent
to Kimball's (1947) hypothesis of variable gene activity). Perhaps
the most attractive working hypothesis is number (3) — catalysis of
homologous templates by the final protein. It is almost identical
with Spiegelman's (1956) hypothesis of adaptive enzyme formation.
He assumes that the newly synthesized enzyme acts autocatalytically
by combining with ribonucleic acid enzyme, forming templates. It
is necessary also to assume that there is competition for the amino
acid pool, and, to explain the mixed character of heterozygotes, that
the antigens catalyze templates produced by all alleles at the locus
in question but not those produced by other loci.

Beale (1957) has emphasized the difficulty in finding a simple
hypothesis to explain all the facts, but the situation seems complex
and it is likely that special assumptions will have to be made in any
case. The point of major significance to be made here is that future
advances in our understanding of the mechanism of serotype determination in *Paramecium* are probably to be gained by the biochemical analysis of the synthetic pathways involved. At present
some mechanism of alternative steady states seems the most likely
explanation, simply because no other reasonable hypothesis remains.

Conclusions and Summary

A large proportion of the stable intraclonal variations in protozoa
do not appear to result from the classic mechanisms of gene mutation, chromosomal aberrations, or changes in ploidy. Attention has
been directed to several types of change associated with the life
cycle in ciliates, the nuclear determinations controlling mating type
in ciliates, plastid variation in *Euglena,* the killer character in *Paramecium,* body form in *Paramecium,* and ciliary antigens in *Paramecium.* Two possible mechanisms for such variations have been
considered: self-reproducing cytoplasmic particles and alternative
steady-state reactions.

The apparently self-reproducing bodies — plastids, kinetosomes,
Feulgen positive bodies, the gullet rudiment in ciliates, and mitochondria — form a diverse group. They are all highly specialized
structurally and, except for the mitochondria and kinetosomes, they

have a highly specialized phylogenetic distribution. Except for the Feulgen positive bodies and possibly plastids, the evidence that they play a genetic role is very poor.

Hypotheses of alternative steady-state reactions suffer from the facts that specific instances are largely unknown and that such hypotheses are too readily adapted to the explanation of almost all sorts of phenomena. Alternative steady states could theoretically exist in the nucleus, the cytoplasm, or even in the extracellular environment. Self-duplicating bodies might even be conceived as special cases of alternative steady-state systems. Furthermore, proof for the existence of such systems would appear to be possible only when the biochemical steps have been worked out in detail. Their frequency, their generality, and their mechanisms are unknown.

Nevertheless, the first clear case has been demonstrated by Cohn (1956), and Spiegelman (1956) has presented evidence in support of his postulate of such a mechanism in adaptive galactozymase formation in yeast. The genetics of mating type substances in *Paramecium* (thought to be protein by Metz, 1954) as well as the inheritance of the immobilization antigens appear to fit no other hypotheses. Alternative steady states adequately explain the other cases of cytoplasmic inheritance for which self-reproducing particles have not been found. Perhaps, as suggested by Spiegelman, there is an autocatalytic step in protein synthesis. It is significant that autocatalysis in protein synthesis was postulated by Waddington (1948) as the essential element in a steady-state theory to explain cellular differentiation in higher organisms.

References

Beale, G. H. 1952. Antigen variation in *P. aurelia,* variety 1. *Genetics* 37: 62–74.

Beale, G. H. 1954. *The Genetics of* Paramecium aurelia. Cambridge University Press, Cambridge.

Beale, G. H. 1957. The antigen system of *Paramecium aurelia. Internat. Rev. Cytol.* 6: 1–23.

Beale, G. H., and H. Kacser. 1957. Studies on the antigens of *Paramecium aurelia* with the aid of fluorescent antibodies. *J. Gen. Microbiol.* 17: 68–74.

Cohn, M. 1956. On the inhibition by glucose of the induced synthesis of β-galactosidase in *Escherichia coli.* In O. H. Goebler (ed.). *Enzymes: Units of Biological Structure and Function.* Academic Press, Inc., New York. Pp. 41–48.

Delbruck, M. 1949. See discussion in Sonneborn, T. M., and G. H. Beale. 1949. Influence des gènes, des plasmagènes, et du milieu dans le déterminisme des caractères antigéniques chez *Paramecium aurelia* variété 4. In *Unités biologiques douées de continuité génétique.* C. N. R. S., Paris 7: 25–36.

Dippell, R. V. 1950. Mutation of the killer cytoplasmic factor in *Paramecium aurelia. Heredity* 4: 165–188.

DIPPELL, R. V. 1955. Some cytological aspects of aging in variety 4 of *Paramecium aurelia*. *J. Protozool.* 2 (Suppl.): 7.

EPHRUSSI, B. 1953. *Nucleo-cytoplasmic Relations in Microorganisms*. Clarendon Press, Oxford.

FAURÉ-FREMIET, E. 1952. Symbiontes bacteriens des ciliés du genre *Euplotes*. *Compt. rend. Acad. Sci.* 235: 402–403.

FINGER, I. 1956. Immobilizing and precipitating antigens of *Paramecium*. *Biol. Bull.* 111: 358–363.

FINGER, I. 1957a. Immunological studies of the immobilization antigens of *Paramecium aurelia*, variety 2. *J. Gen. Microbiol.* 16: 350–359.

FINGER, I. 1957b. The inheritance of the immobilization antigens of *Paramecium aurelia*, variety 2. *J. Genetics* 55: 361–374.

HANSON, E. 1954. Studies on kappa-like particles in sensitives of *Paramecium aurelia*, variety 4. *Genetics* 39: 229–239.

HANSON, E. 1955. Inheritance and regeneration of cytoplasmic damage in *Paramecium aurelia*. *Proc. Nat. Acad. Sci. Wash.* 41: 783–786.

JENNINGS, H. S. 1944. *Paramecium bursaria*: life history. V. Some relations of external conditions, past or present, to aging and to mortality of exconjugants, with summary of conclusions on age and death. *J. Exp. Zool.* 99: 15–31.

KIMBALL, R. F. 1947. The induction of inheritable modification in reaction to antiserum in *Paramecium aurelia*. *Genetics* 32: 486–499.

KIMBALL, R. F., and N. GAITHER. 1954. Lack of an effect of a high dose of X-rays on aging in *Paramecium aurelia*, variety 1. *Genetics* 39: 977.

LWOFF, A. 1950. *Morphogenesis in Ciliates*. John Wiley & Sons, Inc., New York.

MARGOLIN, P. 1954. A method for obtaining amacronucleated animals in *Paramecium aurelia*. *J. Protozool.* 1: 174–177.

METZ, C. B. 1954. Mating substances and the physiology of fertilization in ciliates. In D. H. Weinrich (ed.). *Sex in Microorganisms*. A.A.A.S. Symp. Wash. Pp. 284–334.

MITCHELL, M. B., and H. K. MITCHELL. 1952. A case of "maternal" inheritance in *Neurospora crassa*. *Proc. Nat. Acad. Sci.* 38: 442–449.

MITCHELL, M. B., H. K. MITCHELL, and A. TISSIÈRES. 1953. Mendelian and non-Mendelian factors affecting the cytochrome system in *Neurospora crassa*. *Proc. Nat. Acad. Sci.* 39: 606–613.

MONOD, J. 1947. The phenomenon of enzymatic adaptation and its bearings on problems of genetics and cellular differentiation. *Growth* 11: 223–289.

NANNEY, D. L. 1956. Caryonidal inheritance and nuclear differentiation. *Am. Nat.* 90: 291–307.

NANNEY, D. L., and P. A. CAUGHEY. 1953. Mating type determination in *Tetrahymena pyriformis*. *Proc. Nat. Acad. Sci. Wash.* 39: 1057–1063.

OUDIN, J. 1952. Specific precipitation in gels and its application to immunochemical analysis. *Methods Med. Res.* 5: 335–378.

PIEKARSKI, G. 1949. Blepharoplast and Trypaflavinwirkung bei *Trypanosoma brucei*. *Zentralbl. f. Bakt.* Abt. 1, 153: 109–115.

PREER, J. R., JR. 1950. Microscopically visible bodies in the cytoplasm of the 'killer' strains of *Paramecium aurelia*. *Genetics* 35: 344–362.

PREER, J. R., JR. 1956. A quantitative study of a technique of double diffusion in agar. *J. Immunol.* 77: 52–60.

PREER, J. R., JR., and L. B. PREER. 1958. Gel diffusion studies on the antigens of isolated cellular components in *Paramecium*. *J. Protozool.* (In press).

PRINGLE, C. R. 1956. Antigenic variation in *Paramecium aurelia*, variety 9. *Z. ind. Abstam. Vererbungsl.* 87: 421–430.

PRINGSHEIM, E. G., and O. PRINGSHEIM. 1952. Experimental elimination of chromatophores and eye-spot in *Euglena gracilis*. *New Phytol.* 51: 65–76.

PROVASOLI, L., S. H. HUTNER, and I. J. PINTNER. 1951. Destruction of chloroplasts by streptomycin. *Cold Spring Harbor Symp. Quant. Biol.* 16: 113–120.

REISNER, A. 1955. A method of obtaining specific serotype mutants in *Paramecium aurelia*, stock 169, variety 4. *Genetics* 40: 591–592.

RHOADES, M. M. 1947. Plastid mutations. *Cold Spring Harbor Symp. Quant. Biol.* *11:* 202–207.

SIEGEL, R. W. 1953. A genetic analysis of the mate-killing trait in *Paramecium aurelia,* variety 8. *Genetics 38:* 550–560.

SONNEBORN, T. M. 1937. Sex, sex inheritance and sex determination in *Paramecium aurelia. Proc. Nat. Acad. Sci. Wash. 23:* 378–385.

SONNEBORN, T. M. 1943. Gene and cytoplasm. I. The determination and inheritance of the killer character in variety 4 of *Paramecium aurelia.* II. The bearing of determination and inheritance of characters in *Paramecium aurelia* on problems of cytoplasmic inheritance, pneumococcus transformations, mutations and development. *Proc. Nat. Acad. Sci. Wash. 29:* 329–343.

SONNEBORN, T. M. 1947. Developmental mechanisms in *Paramecium. Growth 11:* 291–307.

SONNEBORN, T. M. 1948. The determination of hereditary antigenic differences in genically identical *Paramecium* cells. *Proc. Nat. Acad. Sci. Wash. 34:* 413–418.

SONNEBORN, T. M. 1950a. Methods in the general biology and genetics of *Paramecium aurelia. J. Exp. Zool. 113:* 87–148.

SONNEBORN, T. M. 1950b. Cellular transformations. *Harvey Lect. 44:* 145–164.

SONNEBORN, T. M. 1954a. Patterns of nucleo-cytoplasmic integration in *Paramecium. Caryologia* (Suppl.) *6,* Pt. 1: 307–325.

SONNEBORN, T. M. 1954b. The relation of autogamy to senescence and rejuvenescence in *Paramecium aurelia. J. Protozool. 1:* 38–53.

SONNEBORN, T. M., and MYRTLE V. SCHNELLER. 1955a. Genetic consequences of aging in variety 4 of *Paramecium aurelia. Genetics 40:* 596.

SONNEBORN, T. M., and MYRTLE V. SCHNELLER. 1955b. Are there cumulative effects of parental age transmissible through sexual reproduction in variety 4 of *Paramecium aurelia? J. Protozool. 2* (Suppl.): 6–7.

SONNEBORN, T. M., MYRTLE V. SCHNELLER, and M. F. CRAIG. 1956. The basis of variation in phenotype of gene-controlled traits in heterozygotes of *Paramecium aurelia. J. Protozool. 3* (Suppl.): 8.

SPIEGELMAN, S. 1956. On the nature of the enzyme forming system. In O. H. Goebler (ed.). *Enzymes: Units of Biological Structure and Function,* Academic Press, Inc., New York. Pp. 67–89.

SPIEGELMAN, S. 1957. Nucleic acids and the synthesis of proteins. In W. D. McElroy and B. Glass (eds.). *The Chemical Basis of Heredity.* The Johns Hopkins Press, Baltimore. Pp. 232–267.

WADDINGTON, C. H. 1948. The genetic control of development. *Symp. Soc. Exp. Biol. 2:* 145–154.

WAGTENDONK, W. J. VAN, B. VAN TIJN, R. LITMAN, A. REISNER, and M. L. YOUNG. 1956. The surface antigens of *Paramecium aurelia. J. Gen. Microbiol. 15:* 617–619.

WILSON, E. B. 1925. *The Cell in Development and Heredity.* The Macmillan Company, New York.

2

Quantitative Chromosomal Changes and Differentiation in Plants

Carl R. Partanen[1]

We are concerned here with two separate phenomena, quantitative chromosomal change and cellular differentiation, which may or may not be related. This is an area that has been the subject of considerable discussion in the past, since certain associations may, in some cases, seem quite obvious. The intent of the present consideration will be to explore this area quite critically, especially in light of some of the more recent work. However, since the area included under the title is quite broad, much of the subject will be treated in generalities, to the extent that it seems justified. Since the frame of reference for the cytological picture will be cellular differentiation, it would be well to consider that first. Even a glance at the literature is sufficient to point out the advisability of trying to define one's working concept of differentiation at the very outset.

Cellular Differentiation

All would probably be in agreement with the statement that cellular differentiation is a process, more specifically a process of cell change, and further that it is a multifarious process. Actually it is the sum total of numerous processes within a cell, each of which may conceivably proceed in different directions and at various rates, to

[1] The Biological Laboratories, Harvard University, Cambridge, Mass. Original work reported here was done partly while a Fellow in Cancer Research of the American Cancer Society, and was supported in part by a grant-in-aid to Professors R. H. Wetmore and T. A. Steeves from the American Cancer Society upon recommendation of the Committee on Growth of the National Research Council.
Present address: Children's Cancer Research Foundation, Boston 15, Massachusetts.

the extent that they are independent of one another; the total of all this would be what we generally recognize as differentiation.

If then we start with a small group of undifferentiated cells (which may of necessity have to remain hypothetical), or perhaps from a state of least differentiation, and proceed to a recognizable end-point, we shall have followed a continuum of differentiation. Every cell that has in any way departed from the state of least differentiation has "differentiated" to a degree. However, any distinction concerning the degree of differentiation must necessarily be arbitrary because the very complexity of the numerous processes involved precludes any quantitative expression of differentiation. Nor are various types of differentiation quantitatively comparable. But one can recognize that there are all types and degrees of differentiation.

In light of results in both animal and plant tissue culture (e.g., see Gautheret, 1942; Trinkaus, 1956), the fixity or stability of cell type as the criterion of differentiation seems untenable and would not enter into the present broad definition in any case. To pursue that line of thought a bit further, fixity of cell type can presumably be arrived at through either of two general types of differentiation. One would be through the loss of something that the cell or cell line originally had and that was essential to its totipotentiality. The other would be through the acquisition or elaboration of something that limits the potential of a cell. To restrict the term differentiation to such cases alone would be to use it in a very narrow and arbitrary sense, and at the cost of a useful concept. From the point of view of the present consideration, processes of differentiation led such cells to the point of no return, but the very first molecular event that initiated a single train of events (of the many involved) constituted differentiation, though at that point of the very lowest degree. If another term is needed, let it be applied to that degree and type of differentiation which (at least to the extent of present knowledge) is fixed or irreversible. This must, however, be a very loose category, since there would be little else than negative evidence to support placing a cell into it.

Consider the heterogeneity of cell types comprising the stem of a vascular plant. One might wonder how many of such cells would fit the criterion of fixity. Obviously, any cell devoid of an essential such as a nucleus can be assumed to be irreversibly differentiated.

Conceivably there are other less obvious irreplaceable deficiencies. It is also reasonable that there is a point in the series of events leading to the total loss of the nucleus after which the changes become irreversible. As for many of the other cells, however, the only deficiency may well be our inability to present such a cell with the conditions in which it can once again express some of its potentialities. In plants the idea of totipotentiality of cells is not merely a theoretical consideration. The regeneration of shoots from callus either *in vivo* or *in vitro* is a common occurrence (e.g., see Priestley and Swingle, 1929; Skoog and Tsui, 1951). Although there are no known cases of entire plants having been grown from single isolated cells *in vitro,* recent reports indicate that the techniques for doing so may soon be at hand (Muir *et al.,* 1954; Riker, 1957).

In the intact plant, there are also factors external to the protoplast that limit the expression of its potentialities. For example, the cell wall, elaborated through the activity of the protoplast itself, becomes a definite limiting factor. The protoplast, in effect, builds a heavy wooden box around itself. Even then, however, the protoplast may occasionally show at least a limited ability to continue, as seen, for example, in the case of septate fibers. These are elongate cells, which, after undergoing a heavy deposition of secondary wall, characteristically undergo further transverse divisions of the protoplast, although there is no escape from the confinement (Vestal and Vestal, 1940). Since mere division is far from an expression of totipotentiality, one can only speculate as to the potentialities of the protoplasts were they not confined.

Another example can be seen in the case of parenchyma cells, which are hemmed in but adjoin either intercellular spaces or tracheary elements. The protoplasts may pass through pits to reach the open area and, once there, they grow and divide to fill all available space, forming tyloses (Gerry, 1914). Again the restraints on the protoplast appear to be largely physical.

Also, consider the vascular cambium. The cambial initials are to a large degree composed of very elongate cells that continue to divide and give off derivatives in two directions, namely, inward to form the xylem and outward, the phloem. However, if one relieves the physical pressures and also unavoidably alters the chemical environment, as, for example, by removing a section of bark, by cutting the stem (Priestley and Swingle, 1929), or by explanting

the tissue *in vitro* (Gautheret, 1942), the elongate character of the
cells is lost and the cell population soon contains only relatively
isodiametric cells. These do not become xylem and phloem with
regularity although they may produce vascular elements more or
less at random. However, if upon such a callus mass growing *in
vitro* one imposes a simple chemical gradient by introducing auxin
locally on top, one gets a much more organized and greater development of vascular tissue, the position and nature of the cells depending upon the concentration of auxin introduced (Wetmore and
Sorokin, 1955).

To place this type of phenomenon on a purely mechanistic and
greatly oversimplified basis, consider a single cell capable of division. No cell exists in an environment such that certain factors, as,
for example, light and gravity, are not acting differentially upon its
different parts. Thus the first plane of division may be determined.
Divisions continue, similarly determined, and the cell mass may
become three-dimensional. Differences between cells become
greater, some being within the mass, others peripheral; thus such
factors as pressure, oxygen tension, light, available substrates or
water, and many other factors may vary, each creating its own
gradient in its own direction, thus providing an almost infinite possibility of slightly different cell environments within the cell mass.
Assume that the genome has reproduced accurately throughout.
If the response of a cell to each slight nuance of environment were
different, the result would probably be utter chaos. However, it is
most likely that a whole range of differences within certain limits
elicits essentially the same response from a genotype. In such a
system, the last or most obvious condition fulfilled to elicit a certain
visible response becomes what has been termed "the initiating
stimulus," as, for example, the auxin on the callus to produce vascular tissue.

This concept of differentiation involves the whole organism, or
at least the organ, as a background; the cell does not become differentiated alone. Thus, differentiation is a function of an individual
cell as a part of a whole, and a plant cell is what it is largely because
of where it is. The cell may display fixity in the situation in which
it has been differentiated; alter the whole and it is no longer the
same population in which the cells became differentiated, and cells
of various types and degrees of differentiation may develop in other
directions.

Quantitative Chromosomal Changes in Normal Cellular Differentiation

Turning our attention now to the other phase of this discussion, namely, quantitative chromosomal changes and their possible relationship to the processes of differentiation, we shall first take a brief general view of the literature. Since the area of somatic chromosome numbers in plants has been reviewed quite recently (D'Amato, 1952; Geitler, 1953), the list of references cited here will by no means be complete. For a much more extensive treatment of the literature, these reviews should be consulted.

By now the occurrence of somatic chromosomal variability in numerous species of plants is well documented. By far the most commonly observed phenomenon is that of somatic polyploidy. From as early as 50 years ago (see D'Amato, 1952), there have been sporadic reports of polyploid mitoses in various tissues of diploid plants. These observations have been made in various ways. For example, it has long been noted that the callus regeneration of a cut stem may have polyploid cells in it and that these may develop into polyploid shoots (Lindstrom and Koos, 1931). Polyploid mitoses were also observed in such phenomena as the formation of root nodules in legumes (Wipf and Cooper, 1940).

The classic material for somatic polyploidy, however, is spinach. In this case polyploidy is a common occurrence in the dividing cells of the root and for that reason was reported as early as 1911 (Stomps, 1911), though it was not fully understood until later (Gentcheff and Gustafsson, 1939; Berger, 1941). Certain other groups of plants are also characterized by regular divisions of polyploid cells among the more common diploids, some of these apparently being limited to the seedling roots, whereas in others the phenomenon persists into later stages (Tjio, 1948; D'Amato, 1952).

Much more commonly, however, the polyploids in such a mixed cell population do not divide again and hence are not as readily detected. Earlier observers tried to make use of various criteria such as nuclear size or the number of heterochromatic bodies as approximations of the degree of ploidy (Huskins and Steinitz, 1948a). However, it was not until the discovery that polyploid mitoses followed treatment with high concentrations of growth hormones (Levan, 1939) and the later realization that these polyploids very likely were already present in the tissues (Huskins and Steinitz, 1948b) that the concept of somatic chromosomal inconstancy became more nearly universal and investigators began to look for it everywhere.

By now polyploidy has been observed in almost every conceivable tissue but by no means in all species, since many species do not show any obvious signs of somatic polyploidy. There are numerous reports of its occurrence in roots and stems, mostly in more mature tissues, though if one's definition of meristem includes anything that happens to be dividing or potentially capable of it, then polyploidy can also be found in the meristems of some plants. However, it does not seem to occur naturally in the meristem of the more limited sense, i.e., that part which is actively or potentially the source of new cells and which is self-perpetuating, a group of cells which might more appropriately be termed "the initials" to distinguish it from the more general meristematic zone. That, of course, is entirely expected, since these cells are potentially in what might be loosely referred to as the germ line.

In addition to roots and stems, polyploidy has been observed in floral and fruiting structures; for example, all sorts of modifications of the polyploid theme occur in the anther tapetum, ranging from polyploidy to multiple nuclei per cell (see D'Amato, 1952). Also, in the angiosperm embryo sac, polyploidization, other than that which is a regular process, has been observed, both in synergid and antipodal nuclei (Beaudry, 1951; Håkansson, 1951). The angiosperm endosperm, which in many cases is already polyploid with respect to the embryo, may also undergo further polyploidization (Duncan and Ross, 1950; Punnett, 1953; Straus, 1954). Polyploidy has also been observed in tissues of fruits (see D'Amato, 1952). The only additional observation of somatic polyploidy that I might add to this cataloging is its occasional occurrence in the gametophytic (haploid) generation of the ferns (Partanen, 1959a).

So far we have made no mention of the mechanisms by which this somatic polyploidy comes about. All of the somatic polyploidy in plants consists essentially of a range of variations on a single theme, the theme being that during the interphase not one but two or more replications of the chromosomes can occur. Thus, if and when the nucleus does come to division, there are more than two units to be separated. This process of additional replication has been referred to as either *endomitosis* or *endoreduplication*, some authors preferring the latter term (Levan and Hauschka, 1953) to distinguish it from "endomitosis" as it was first applied by Geitler (1939), which included detectable signs of the process during the interphase, whereas "endoreduplication" has been applied to that in which there are no such evidences. However, they are probably all one and the same process, the main feature being that no spindle is formed after the replication is completed; thus what would have been daughter chromosomes in separate nuclei remain instead in close proximity in the same nucleus and the replication is once again initiated, both units being duplicated. When such a nucleus, which has undergone one additional replication, comes to division, the diagnostic characteristic is the paired appearance of the chromosomes, since they are still in close proximity. If they are still attached at the centromere, these are called *diplochromosomes*. If two additional replications occur during the interphase, the chromosomes are quadripartite and are called *quadruplochromosomes*. Again there is considerable variation among species, and sometimes within individuals, in the degree of separation of the multiple chromosome units as division begins. They may range from those that are completely separated but near one another in the earliest stages of the mitotic division, through various degrees of attachment and relational coiling that may persist to the metaphase.

Thus far we have discussed only increases in chromosome number. Obviously another kind of quantitative change is also possible, namely, a decrease. Decreases of a very precise manner are of course characteristic of meiosis, but we are concerned here only with somatic cells. Certainly reductional divisions in plant somatic cells are a rarity. However, this does not rule out their possible occurrence. Considering the relatively brief duration of mitosis, the chances of ever seeing a very rare type of division are exceedingly small. But, accepting the observations that apparent reductions of chromosome number can be effected by chemical treatments, one must allow the possibility that the mechanism (or specific misfunction of a mechanism) exists and that it might also operate under natural conditions. The analysis by Patau and Patil (1951) of the so-called "somatic reductions" induced by treatments with a nucleic acid salt indicate that the total picture is one of a general effect upon the cell and that many of the observed cytological deviations can be explained on the basis of a disturbed spindle mechanism. Similar results have been observed by others, using other substances (see D'Amato, 1952). In all of these observations there appears, however, to be no real evidence for a regular preferential distribution of chromosomes; but this does not exclude the possibility of such preferential distributions. They may even be characteristic of certain special situations (for further discussion of such phenomena, see, e.g., Huskins, 1948; Patau and Patil, 1951; Gläss, 1957). However, it would be extremely foolhardy to make general application of isolated observations. This does emphasize once again a point of which certainly most cytologists have long been aware, that in the nucleus many things which ordinarily do not happen may be possible. That is, if we were to understand the nucleus completely, we could perhaps list occurrences which are possible, mechanically and otherwise. However, lacking that vantage point of omniscience, we can but acknowledge the obvious; that which can be shown unquestionably to occur must be admitted to the realm of possibility even though we do not understand the special conditions that had to prevail to bring about this departure from the usual. That there are to be found numerous examples of various types of apparent deviations from the usual pattern of events indicates of course the complexity of the processes we are studying. They should not, however, be permitted to invalidate completely our attempts to generalize on the more common situation to the extent that we are able to understand it. The exceptions, of great value in analyzing the mechanisms involved, must of course be noted but should be kept in their proper perspective. On the basis of the present evidence, then, one must conclude that somatic reductional divisions, although they may occur occasionally, are probably of no real significance in the differentiation processes. A similar type of reduction, which has been observed in some neoplasms, is the multipolar division, i.e., a division with more than two poles to the spindle. This seems also to result in a random distribution of chromosomes.

So far our evidence on quantitative nuclear changes has consisted of visible changes, i.e., chromosome number. This is quite satisfactory for those cells that divide, but one would especially like to know about the nondividing cells. Nuclear size is not a good criterion of degree of ploidy, since nuclei obviously contain some variable components that vary independently of ploidy. However, DNA appears to be quite constant per chromatid set, thus behaving as one might expect genic material to behave. Through a number of fortunate circumstances and despite certain a priori theoretical objections, it has by now been quite well established on empirical grounds that microspectrophotometric measurement of the Feulgen reaction, i.e., the amount of bound dye in individual nuclei, is proportional to the amount of DNA in such nuclei (see Swift, 1953; Leuchtenberger, 1954; Pollister and Ornstein, 1955). There has further developed the very good working hypothesis that the amount of DNA per chromatid set is more or less constant (Swift, 1950, 1953; Pollister et al., 1951; Alfert and Swift, 1953; Patau and Swift, 1953). During the interval between telophase and the following prophase the amount of DNA per nucleus doubles. Intermediate DNA-Feulgen values obtained during this time are generally taken to be true intermediates in the synthesis of DNA. Thus, by measuring the Feulgen reaction in individual nuclei, one can determine their level of ploidy quite readily. Refined techniques have increased the versatility of the technique so that through minimizing distributional error, one can also measure irregular nuclei and division figures with a reasonably high degree of accuracy (Ornstein, 1952; Patau, 1952). However, the technique is not infallible; there are various possible sources of error. Therefore the technical aspects must always be carefully assayed before attempting to interpret results.

In brief, this technique has served to confirm what had already been demonstrated to a lesser extent by other means. DNA studies have shown polyploid classes in the expected places, and have extended those observations by revealing higher classes in many cases (see Swift, 1953). While some workers have found apparent intermediate DNA levels to be characteristic of certain tissues (Schrader and Leuchtenberger, 1949; Bryan, 1951), others have found, upon reinvestigation, that such results were probably the result of technical error and they have presented considerable data to support the constancy hypothesis (Swift, 1950; Moses and Taylor, 1955; Woodard, 1956). This technique, with the proper precautions, is a valuable one and seems to promise much more, especially if coupled with other techniques such as autoradiography, as has been done in the nice DNA stories of meiosis in *Lilium* and *Tradescantia* (Taylor and McMaster, 1954; Moses and Taylor, 1955).

A consideration of quantitative chromosomal changes must also recognize that there are other components of nuclei besides DNA which may well play a role in such phenomena as differentiation. For example, there are proteins, which comprise by far the major part of the dry weight of nuclei, and RNA. Unfortunately the quantitative cytochemistry of these substances is far less certain than that of DNA. Nevertheless, there are some suggestions, for example, of possible roles of histone in the modification of genic expression (see Bloch and Godman, 1955). However, such considerations are beyond the intended scope of the present discussion.

Polyploidy in Plant Neoplasia and Tissue Culture

The relationship of polyploidy to neoplastic growth has long been an area for conjecture. Just as in animal tumors, its occurrence in plant tumors is not universal. However, when it does occur, it comes about by the same mechanism that we have already discussed. Therefore, the subject is quite relevant to our present discussion, and we shall now become a bit more specific by considering one situation in which it has been studied in some detail.

Ideally, to approach the question of the relationship of polyploidy to tumorization, one should be able to follow a tumor cytologically from its very beginning. Unfortunately, there is the very real problem of recognizing a tumor (or what will become a tumor) at such an early stage. This is especially true when the tumor occurs in a complex organism. The ideal situation can at least be approximated by studying tumors that occur in relatively simple organisms, especially if they arise from known cell types, grown in sterile culture *in vitro* on completely defined media. The studies that I shall now describe were carried out under these conditions.

By way of orientation to the subject under investigation let us consider briefly the life cycle of a fern. The fern plant as seen in the field is the diploid sporophyte generation. It forms spores that are haploid, since meiosis occurred in their development. These spores germinate to form separate small, relatively simple plants, which of course are also haploid. These prothalli (gametophytes) produce eggs and sperm which upon fusion once again initiate the diploid generation.

However, our interest at present is in the haploid generation, which is entirely separate from the sporophyte and which is a full-fledged organism in its own right, quite capable of continuing indefinitely under proper conditions. The extent of its potentialities becomes more evident when it is cultured aseptically *in vitro.* Spores are sown on a simple mineral solution with agar and in time they develop into prothalli, which are essentially sheets of cells, one cell thick, with additional layers centrally and with rhizoids. Habit photographs of prothalli are shown on Plate I (1) and (2).

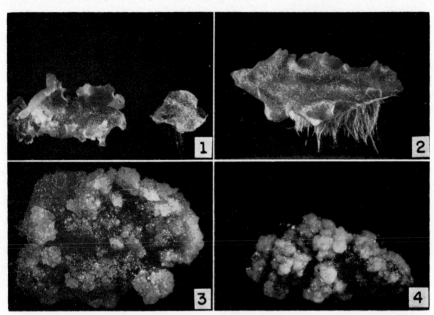

Plate I. Habit photographs of fern prothalli (above) grown aseptically *in vitro* and their corresponding tumors (below). (1) Young prothallus of *Pteridium aquilinum* (right) showing the classic "heart" shape. Older prothallus (left), which is more typical of those grown *in vitro* for longer periods of time, showing vegetative multiplication in the form of new prothalli arising as marginal outgrowths, especially along the left and lower edges, to form a clone of prothalli. (2) Characteristic prothallus of *Osmunda cinnamomea* showing large numbers of rhizoids arising centrally on the lower side. (3) Tumor isolated from prothallial culture of *Pteridium* and grown in separate culture. Most of the tumors have this characteristically nodular appearance through local centers of activity. (4) Tumor from prothallus of *Osmunda cinnamomea*. This strain has now been in continued culture for over eight years.

These prothalli are potentially unlimited in their growth, since they continue to increase vegetatively. Unlimited quantities of clonal haploid plants are thus readily available.

In the process of culturing the prothalli of *Pteridium aquilinum* (the bracken fern) *in vitro,* occasional tumorlike cell masses are encountered, which, when isolated, retain their characteristic appearance and continue to grow in a completely unorganized manner

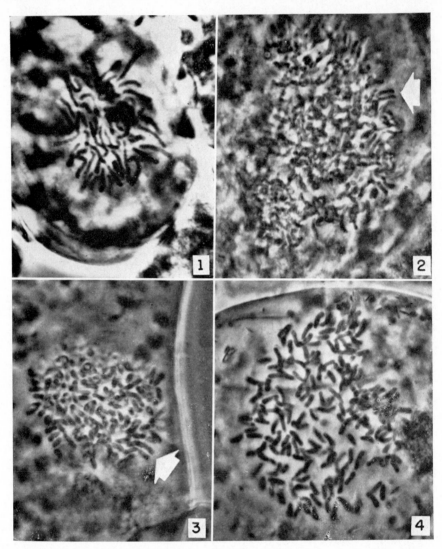

Plate II. Series of nuclear figures from prothallus and tumors of *Pteridium aquilinum* showing evidences of the mechanism of chromosome number increase through endoreduplication. (1) Normal haploid set of 52 chromosomes in prothallus. (2), (3) Figures found in tumor derived from prothallus of *Pteridium*. Paired appearance of sister chromosomes (e. g., see arrows) in nuclei approaching metaphase indicates the occurrence of an additional replication of chromosomes during the preceding interphase. (4) The result of the foregoing process, showing a high number of chromosomes, in the 4N range. Mag. ca. × 2260 in all parts of Plate. (From Partanen *et al.*, 1955; *courtesy, American Journal of Botany*.)

in separate culture (Steeves *et al.*, 1955). Typical fern gameto-phyte tumors are shown on Plate I (3) and (4).

Cytological studies on these tumors (Partanen *et al.*, 1955) showed that initially all of the divisions that occur in them are haploid, and the nuclear condition appears to be normal in all re-spects. In subsequent culture, however, the chromosome numbers in the dividing cells become higher, usually reaching a level around tetraploid. This comes about through endoreduplication, as can be seen on Plate II.

Superimposed upon this polyploidy, and very prevalent, is a variable aneuploidy within single cultures. Despite these nuclear changes, the tumors appear to be able to continue growing in-definitely. Occasionally there occur local reversions to a more nor-mal prothallial expression within such tumors and these have had lower, though not haploid, chromosome numbers.

Against this cytological background, these fern prothalli and tumors were studied through microphotometric measurement of the Feulgen reaction to estimate the amounts of DNA in individual nuclei (Partanen, 1956). The results of this study are shown in Fig. 6. The relative amounts of DNA in individual nuclei are plotted on the horizontal axis and the numbers of nuclei on the vertical.

The normal prothallus is at the top. The graph shows one peak of DNA values that trails off to approximately twice this amount. This shows the amount of variation in DNA per nucleus in normal prothalli of *Pteridium aquilinum*, and can be taken to represent synthesis of DNA from the unduplicated level to the duplicated level. It would seem that since there is no second peak at the dupli-cated level, division follows closely upon completion of DNA syn-thesis.

In striking contrast with this is the situation in the older strains of tumor, in which the chromosome numbers had already increased (as shown in the bottom two graphs). These are characterized by a whole spectrum of DNA values. If one measured enough nuclei, the whole scale would probably be blocked in, since there are ac-tually two sources of DNA variability here: synthesis of DNA and the variable aneuploidy. The higher values were of course entirely predictable, i.e., if chromosome numbers were in the 3N to 4N range, one would expect DNA values to range from the undupli-cated level, 3C to 4C, on up to around 8C, which would be the duplicated 4N level. The presence of even higher levels indicates continued endoreduplication beyond that which was detected by chromosome counting. However, of more interest are the lower values, some of which actually reach down into the normal 1C to

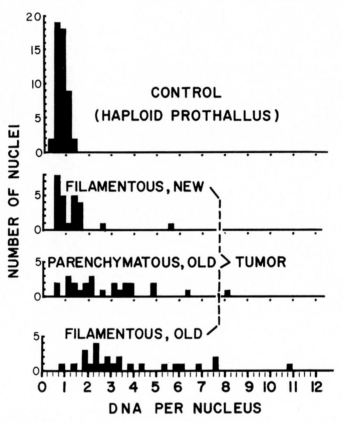

Fig. 6. Relative amounts of DNA (Feulgen) in individual nuclei of normal and tumorous growth of prothalli of *Pteridium aquilinum*. Top graph indicates the normal range of DNA values in haploid gametophyte. Second graph shows the beginnings of endomitotic reduplication in a new tumor (isolated 1 month) by the accumulation of DNA values at the duplicated level. Lower two graphs show the extreme range of DNA in nuclei of old tumors, which characteristically show polyploidy and variable aneuploidy. (From Partanen, 1956; *courtesy, Cancer Research.*)

2C range of the prothallus. Whether these are actually nuclei with the haploid number of chromosomes or simply low aneuploids in that vicinity is not known. However, the fact that local reversions to a more normal expression did have lower chromosome numbers is suggestive of the possibility that there may well be cells in these tumors which are still essentially normal cytologically. It is reasonable to assume that other differences also exist, and that the polyploidy may be but one manifestation of such differences. The cells comprising such heterogeneous populations as these tumors may be of greatly varying potentialities. One thing is apparent: the polyploid cells are not necessarily headed down a dead-end street; they may continue to function and divide apparently indefinitely.

We have seen that the normal prothallus has a rather tight group of DNA values, strictly within the haploid limits, while the older tumors have a wide range of values. It is, then, of interest to see a transitional stage from one condition to the other. This is seen in the second graph. This was a new tumor, during its first passage in separate culture, and it showed no change in chromosome number in the divisions seen. Although most of the DNA values fall within the normal range, there is now a second peak at the duplicated (2C) level. This would indicate that quite unlike the nuclei in the normal prothallus, those in the new tumor begin to show a tendency to at least pause at the duplicated level before dividing, and perhaps not divide at all, but instead continue to synthesize more chromosomal material. This would, then, be the beginnings of endoreduplication, much after the tumorous habit had been established, since this tumor had already been recognized and isolated about a month before. Thus, polyploidization quite clearly appears to be a secondary phenomenon in this type of tumorization.

In quite another family of ferns, the gametophytes of *Osmunda cinnamomea* (the cinnamon fern) have also given rise to tumors *in vitro* (Morel and Wetmore, 1951). It is interesting that these tumors have a similar nuclear behavior to those already described in *Pteridium* (Partanen *et al.*, 1955; Partanen, 1956). The DNA histograms (Fig. 7) show first the normal prothallus at the top. Here there is a pronounced second peak; this is quite appropriate, since earlier cytological observations indicated that endoreduplication did occur in this species more frequently than in *Pteridium*, and although this does not show actual endopolyploidy, it does show a tendency in that direction. That is, a separation of the two processes of replication and division is the first condition necessary for endoreduplication.

The extremely variable polyploid tumor is shown at the bottom, and is like those already seen in *Pteridium*.

Of special interest here is the second graph, which shows the DNA distribution in a very old strain of tumor (seven years). It seems quite stable at the approximate diploid level. Earlier cytological studies had shown that this tumor also had higher chromosome numbers, and it is quite certain that its DNA pattern would have been very much like the bottom one at one time. Obviously a tumor has quite an array of cells and chromosome combinations (gene

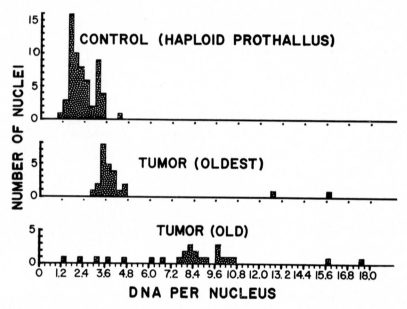

Fig. 7. Relative amounts of DNA (Feulgen) in individual nuclei of prothalli of *Osmunda cinnamomea* and tumors derived from them. Top graph shows the range of values in normal prothallus, with distinct second peak at the duplicated level, indicative of possible tendency toward endoreduplication. Lower two graphs are from tumors, the middle one showing the present state of a previously high polyploid tumor, and the lower one a typically variable polyploid-aneuploid. (From Partanen, 1956; *courtesy, Cancer Research.*)

combinations) to choose from. It would seem very likely that once in a while a really good combination for the prevailing conditions would occur. In any case, one type has clearly dominated in this strain.

After seeing how those two unrelated tumors behaved *in vitro*, the question arose as to how prevalent such nuclear behavior might be in unorganized plant growth in general, in sterile culture. That is, could there be something about the test-tube environment of the growths that is conducive to such changes? It was, therefore, decided to investigate something quite different, under a variety of conditions.

Tubers of *Helianthus tuberosus* (a sunflower) have long been used as a source of plant tissue cultures. In our investigations this turned out to be a fortunate choice, since it presented quite a

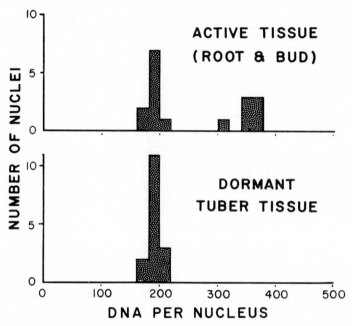

Fig. 8. Relative amounts of DNA (Feulgen) in individual nuclei of *Helianthus tuberosus* tissues, comparing the starting point of tissue cultures (dormant tuber tissue, lower graph) with actively dividing tissues of root and bud. In the active tissue, the second DNA level was identified as the duplicated diploid (4C) level by the measurement of prophases and metaphases, which fell into this group.

different story from the preceding. All measurements in this study were made by the two-wavelength method. Technical aspects and other details are to be reported elsewhere (Partanen, 1959b).

As a point of departure, the top graph in Fig. 8 characterizes the nuclear condition of the dormant tuber tissue from which the tissue cultures are started. This shows a tight grouping of DNA values at the unduplicated level (2C), as can be seen by comparing this with the active tissue seen in the lower graph. The duplicated (4C) level was identified by measuring prophases and metaphases. Thus, our point of departure is a completely diploid tissue throughout; we have yet to find values any higher in this tissue.

The next obvious thing to do was to culture the tissue. As can be seen in Fig. 9, the values all remained in the normal range in the tissue culture. There was no indication of any major shift.

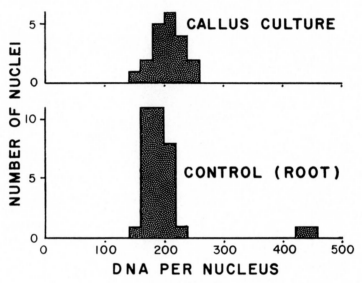

Fig. 9. Relative amounts of DNA (Feulgen) in individual nuclei of *Helianthus tuberosus* tissue cultures as compared with the normal 2C to 4C limits of the root, showing that the tissue cultures remained well within the normal limits.

Helianthus tuberosus is classic material in its response to growth hormone, since it requires auxin for growth *in vitro* and shows varying responses to varying concentrations (Gautheret, 1947). Fig. 10 shows the DNA histograms of tissues placed on a series of concentrations of naphthalene acetic acid. Despite the varied responses, ranging from extreme cell growth with little or no division at the highest concentration, through an optimal concentration for cell division, to no response with no auxin, there were no nuclear changes that were anywhere near the order of magnitude of a single chromosome set. The tissues all remained quite well in the normal range. The slight variation around the 2C level is of the order that one does not ordinarily attempt to explain. It may be real or it may not. In the case of the explant on no auxin, however, the lower values are

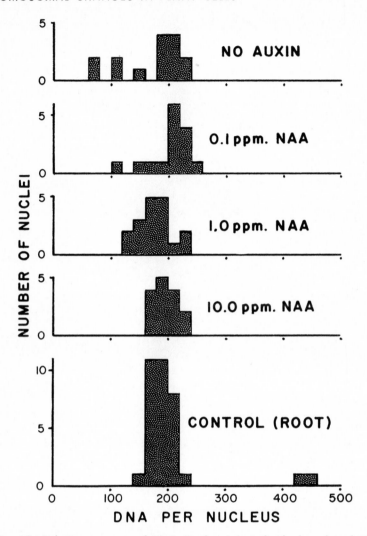

Fig. 10. Relative amounts of DNA (Feulgen) in individual nuclei of *Helianthus tuberosus* tuber tissue explants cultured on a range of concentrations of naphthalene acetic acid (indicated in parts per million) as compared with the normal root tissue. (This experiment was processed and measured with that shown in Fig. 9, hence the grouped controls.) The range of cellular reactions to the auxin series is shown to have no appreciable counterpart at the quantitative nuclear DNA level.

due to obvious pycnosis of the nuclei; the tissue was undoubtedly dying.

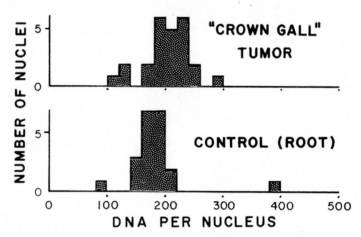

Fig. 11. Relative amounts of DNA (Feulgen) in individual nuclei of "crown gall" induced *in vitro* by inoculating *Helianthus tuberosus* tuber tissue explants with *Agrobacterium tumefaciens*, compared with the 2C to 4C range of the root.

Crown gall tumors can be induced in *Helianthus* by inoculating with a virulent strain of *Agrobacterium tumefaciens*. This was done *in vitro*, and the nuclear condition of the resulting tumorous growth is shown in the top graph of Fig. 11. As could by now perhaps be expected, the nuclei still remained in the normal range. There was again a bit of variability but of a very low order.

Discussion

The results of these studies would seem to have some strong implications in relation to the problem of quantitative chromosomal change and differentiation. Quite obviously, not only from this example, but from countless others, cellular differentiation in plants can proceed very well in the absence of gross quantitative chromosomal changes. Clearly, then, no direct relationship, either causal or consequential, can be made between polyploidy and cellular differentiation in general. However, this is only reasonable and self-evident, if one considers the diversity of phenomena that can be included under the term *differentiation*.

Upon analyzing the total phenomenon of cell multiplication, it becomes obvious that there are several separate and separable processes involved. Since cell division includes nuclear division, a necessary prerequisite for division is of course the replication of new chromosomal material prior to division. It appears that the exact duplicated level of DNA is important for the initiation of nuclear division. The next general process, nuclear division, appears to be essential for the following step, the division of the cell itself. That these processes are separable is obvious from the occurrence of endopolyploidy and binucleate cells.

The problem then hinges on the fact that the difference between the retention of the basic chromosome number and the occurrence of endopolyploidy is the degree to which these sequential processes of replication and mitosis are coupled. In those plants that undergo endoreduplication, obviously the degree of coupling varies locally within the plant. In the meristems proper, the coupling must remain strict because there the characteristic chromosome number is always retained (except in cases of certain environmentally caused mitotic upsets). To do otherwise would be fatal to the species, since they would polyploidize themselves out of existence. Any such tendency, which would most likely be genetically determined, would of course be selected against. Nonmeristematic cells, on the other hand, are not potentially in the germ line (although they may become so through alteration of the existing system); therefore, they probably encounter direct genetic selection only to the extent that they affect the functioning of the organism. Conceivably polyploidy, or some variant of it, may be advantageous in certain situations in which it is fairly common, as, for example, in the anther tapetum.

The important question, then, is "what factors underlie the coupling of chromosomal replication and mitosis?" Although it is often so stated, in the strictest sense it is not true that "DNA is being synthesized for the next division." Many things can happen and division is but one thing that may occur upon the completion of a unit of replication. One alternative is continued replication.

That there are genetic factors underlying the occurrence of endoreduplication can also be inferred from the observation that certain taxonomic groups of plants seem to have a tendency toward it. That it occurs in tumorous growth seems to be but a further expression of a natural tendency. Thus, the occurrence of somatic polyploidy, even to a slight degree, in the normal fern prothallus indicates that the mechanism is there and, for reasons unknown, this becomes enhanced in the tumors. That some tumors display a prevalence of polyploid divisions is understandable if one considers two components of the phenomenon separately. One would be the tendency toward the uncoupling of chromosomal replication and mitosis, leading to endoreduplication. The other would be the predominant characteristic of tumorous growth, namely, the lack of normal restraints on cell division. Thus, in species characterized by endoreduplication, the normal occurrence of polyploidy goes undetected because generally those cells do not divide again. If, however, that part of the cell population does divide, of course the increased chromosome number is revealed. Further, the endopolyploid part of the cell population, by virtue of its continuing divisions in the tumor, continues to increase, thereby making the polyploidy even more prevalent.

The *Helianthus* tissue, on the other hand, seems to show no such tendency *in vivo*, and a number of treatments *in vitro*, including tumorization, did nothing to alter this. Other observations that fit into this pattern are those of Rasch *et al.* (1957), in which they studied the cytological changes in broad bean during tumor genesis. Broad bean is known to undergo endoreduplication in various tissues (Coleman, 1950). The photometric DNA data show that in the usual range of nuclei from 2C to 16C, a large number of the nuclei are at the 4C level (duplicated diploid), and that when these are stimulated to proliferate by wounding, there is a shift to the 2C (unduplicated diploid) level in the wound periderm. In contrast to this normal picture, crown gall tumors very early begin to show a definite shift toward the higher levels of ploidy and in time show a great number of high polyploids, higher than can be found in the normal plant. Here, again, seems to be another example of the type of thing we have already discussed, in which the normal tendency is exaggerated in the tumor.

How are we to look at the problem of polyploidy in relation to differentiation? Thinking in terms of the numerous factors involved in the processes of nuclear replication and division, ranging from the availability of the necessary building blocks to the energy requirements (e.g., Bullough, 1952), it is reasonable to assume that various of these may become limiting in different parts of the plant and these limitations may have a range of manifestations, depending upon how general the mechanism that is affected. These are, again, some of the factors that are undoubtedly involved in differentiation. Thus, of the numerous avenues that cellular differentiation may take, polyploidization is one. It may or may not precede, accompany, or follow other types of changes that also constitute differentiation. It may in some cases be associated with other specific types of differentiation because they may both result, in part, from common mechanisms.

As to the nature of the factors and mechanisms, this conceivably involves the total economy of the cell. The degree of "specificity" of the factors involved would seem to be largely a matter of their proximity to the observable effect in the cellular chains of events. It would seem that these processes are open to possible experimental manipulation, one of the more hopeful lines of approach perhaps being the further use of *in vitro* techniques in which environmental and nutritional factors are subject to much more control than is the natural situation. Cytophotometric techniques furnish the necessary means of detecting such nuclear changes. Then, in having two distinctly opposite types of tissues, the *Helianthus*, which does not seem to undergo endoreduplication, and the fern gametophyte tumor, which undergoes it in quite a predictable manner, it would seem that we have the tools for approaching the problem from both directions.

References

ALFERT, M., and H. SWIFT. 1953. Nuclear DNA constancy: A critical evaluation of some exceptions reported by Lison and Pasteels. *Exp. Cell Research 5:* 455–460.

BEAUDRY, J. R. 1951. Seed development following the mating of *Elymus virginicus* L. × *Agropyron repens* (L.) Beauv. *Genetics 36:* 109–133.

BERGER, C. A. 1941. Reinvestigation of polysomaty in spinach. *Bot. Gaz. 102:* 759–769.

BLOCH, D. P., and G. C. GODMAN. 1955. A microphotometric study of the synthesis of desoxyribonucleic acid and nuclear histone. *J. Biophys. Biochem. Cytol. 1:* 17–28.

BRYAN, J. H. D. 1951. DNA-Protein relations during microsporogenesis of *Tradescantia. Chromosoma 4:* 369–392.

BULLOUGH, W. S. 1952. The energy relations of mitotic activity. *Biol. Rev. 27:* 133–168.

COLEMAN, L. C. 1950. Nuclear conditions in normal stem tissue of *Vicia faba. Canad. J. Research* (C), *28:* 382–391.

D'AMATO, F. 1952. Polyploidy in the differentiation and function of tissues and cells in plants. *Caryologia 4:* 311–358.

DUNCAN, R. E., and J. G. ROSS. 1950. The nucleus in differentiation and development, III. Nuclei of maize endosperm. *J. Hered. 41:* 259–268.

GAUTHERET, R. J. 1942. *Culture des Tissus Vegetaux.* Masson et Cie, Paris.

GAUTHERET, R. J. 1947. Plant tissue culture. *Growth* (Suppl.) *6:* 21–43.

GEITLER, L. 1939. Die Entstehung der polyploiden Somakerne der Heteropteren durch Chromosomenteilung ohne Kernteilung. *Chromosoma 1:* 1–22.

GEITLER, L. 1953. Endomitose und endomitotische Polyploidisierung. *Protoplasma-tologia VI*, C. Springer-Verlag OHG, Vienna.

GENTCHEFF, G., and Å. GUSTAFSSON. 1939. The double chromosome reproduction in *Spinacia* and its causes. *Hereditas 25:* 349–358.

GERRY, ELOISE. 1914. Tyloses: Their occurrence and practical significance in some American woods. *J. Agric. Research 1:* 445–469.

GLÄSS, E. 1957. Das Problem der Genomsonderung in den Mitosen unbehandelter Rattenlebern. *Chromosoma 8:* 468–492.

HÅKANSSON, A. 1951. Parthenogenesis in *Allium*. *Botaniska Notiser* (2): 143–179.

HUSKINS, C. L. 1948. Segregation and reduction in somatic tissues I. Initial observations on *Allium cepa*. *J. Hered. 39:* 311–325.

HUSKINS, C. L., and LOTTI M. STEINITZ. 1948a. The nucleus in differentiation and development. I. Heterochromatic bodies in energic nuclei of *Rhoeo* roots. *J. Hered. 39:* 35–43.

HUSKINS, C. L., and LOTTI M. STEINITZ. 1948b. The nucleus in differentiation and development. II. Induced mitoses in differentiated tissues of *Rhoeo* roots. *J. Hered. 39:* 67–77.

LEUCHTENBERGER, C. 1954. Critical evaluation of Feulgen microspectrophotometry for estimating amount of DNA in cell nuclei. *Science 120:* 1022–1023.

LEVAN, A. 1939. Cytological phenomena connected with the root swelling caused by growth substances. *Hereditas 25:* 87–96.

LEVAN, A., and T. S. HAUSCHKA. 1953. Endomitotic reduplication mechanism in ascites tumors of the mouse. *J. Nat. Cancer Inst. 14:* 1–43.

LINDSTROM, E. W., and K. KOOS. 1931. Cytogenetic investigations of a haploid tomato and its diploid and tetraploid progeny. *Am. J. Bot. 18:* 398–410.

MOREL, G., and R. H. WETMORE. 1951. Fern callus tissue culture. *Am. J. Bot. 38:* 141–143.

MOSES, M. J., and J. H. TAYLOR. 1955. Desoxypentose nucleic acid synthesis during microsporogenesis in *Tradescantia*. *Exp. Cell Research 9:* 474–488.

MUIR, W. H., A. C. HILDEBRANDT, and A. J. RIKER. 1954. Plant tissue cultures produced from single isolated cells. *Science 119:* 877–878.

ORNSTEIN, L. 1952. The distributional error in microspectrophotometry. *Lab. Investigation 1:* 250–262.

PARTANEN, C. R. 1956. Comparative microphotometric determinations of desoxyribonucleic acid in normal and tumorous growth of fern prothalli. *Cancer Research 16:* 300–305.

PARTANEN, C. R. 1959a. Endoreduplication in the gametophytic generation of ferns. (In preparation).

PARTANEN, C. R. 1959b. Microphotometric studies on desoxyribonucleic acid in plant tissues *in vitro*. (In preparation).

PARTANEN, C. R., and T. A. STEEVES. 1956. The production of tumorous abnormalities in fern prothalli by ionizing radiations. *Proc. Nat. Acad. Sci. 42:* 906–909.

PARTANEN, C. R., I. M. SUSSEX, and T. A. STEEVES. 1955. Nuclear behavior in relation to abnormal growth in fern prothalli. *Am. J. Bot. 42:* 245–256.

PATAU, K. 1952. Absorption microphotometry of irregular-shaped objects. *Chromosoma 5:* 341–362.

PATAU, K., and R. K. PATIL. 1951. Mitotic effects of sodium nucleate in root tips of *Rhoeo discolor* Hance. *Chromosoma 4:* 470–502.

PATAU, K., and H. SWIFT. 1953. The DNA content (Feulgen) of nuclei during mitosis in a root tip of onion. *Chromosoma 6:* 149–169.

POLLISTER, A. W., and L. ORNSTEIN. 1955. Cytophotometric analysis in the visible spectrum. In R. C. Mellors (ed.). *Analytical Cytology*, Blakiston Division, McGraw-Hill Book Co., Inc., New York. Pp. 1/3–1/79.

POLLISTER, A. W., H. SWIFT, and M. ALFERT. 1951. Studies on the desoxypentose nucleic acid content of animal nuclei. *J. Cell. Comp. Physiol. 38* (Suppl. 1):101–119.

PRIESTLEY, J. H., and C. F. SWINGLE. 1929. Vegetative propagation. *U.S.D.A. Tech. Bull.* No. 151. 98 pp.

PUNNETT, H. H. 1953. Cytological evidence of hexaploid cells in maize endosperm. *J. Hered. 44:* 257–259.

RASCH, E. M., H. SWIFT, and R. M. KLEIN. 1957. Cytological changes during tumor genesis in broad bean. (In preparation).

RIKER, A. J. 1957. Discussion in Decennial Review Conference on Tissue Culture. Woodstock, Vt., 1956. *J. Nat. Cancer Inst. 19:* 589–590.

SCHRADER, F., and C. LEUCHTENBERGER. 1949. Variation in the amount of desoxyribose nucleic acid in different tissues of *Tradescantia. Proc. Nat. Acad. Sci. 35:* 464–468.

SKOOG, F., and C. TSUI. 1951. Growth substances and the formation of buds in plant tissues. In F. Skoog (ed.). *Plant Growth Substances,* University of Wisconsin Press, Madison. Pp. 263–285.

STEEVES, T. A., I. M. SUSSEX, and C. R. PARTANEN. 1955. *In vitro* studies on abnormal growth of prothalli of the bracken fern. *Am. J. Bot. 42:* 232–245.

STOMPS, T. J. 1911. Kernteilung und Synapsis bei *Spinacia oleracea* L. *Biol. Centralbl. 31:* 257–309.

STRAUS, J. 1954. Maize endosperm tissue grown *in vitro.* II. Morphology and cytology. *Am. J. Bot. 41:* 833–839.

SWIFT, H. 1950. The constancy of desoxyribose nucleic acid in plant nuclei. *Proc. Nat. Acad. Sci. 36:* 643–654.

SWIFT, H. 1953. Quantitative aspects of nuclear nucleoproteins. *Internat. Rev. Cytol. 2:* 1–76.

TAYLOR, J. H., and RACHEL D. McMASTER. 1954. Autoradiographic and microphotometric studies of desoxyribose nucleic acid during microgametogenesis in *Lilium longiflorum. Chromosoma 6:* 489–521.

TJIO, J. H. 1948. The somatic chromosomes of some tropical plants. *Hereditas 34:* 135–146.

TRINKAUS, J. P. 1956. The differentiation of tissue cells. *Am. Naturalist 90:* 273–289.

VESTAL, P. A., and MARY R. VESTAL. 1940. The formation of septa in the fibertracheids of *Hypericum androsaemum* L. *Bot. Mus Leaflets, Harvard Univ. 8:* 168–180.

WETMORE, R. H., and S. SOROKIN. 1955. On the differentiation of xylem. *J. Arnold Arb. 36:* 305–317.

WIPF, L., and D. C. COOPER. 1940. Somatic doubling of chromosomes and nodular infection in certain Leguminosae. *Am. J. Bot. 27:* 821–824.

WOODARD, J. W. 1956. DNA in gametogenesis and embryogeny in *Tradescantia. J. Biophys. Biochem. Cytol. 2:* 765–776.

3

Numerical Variation of Chromosomes in Higher Animals

T. C. Hsu [1]

In the summer of 1956 I wrote a chapter reviewing the subject of "chromosomes of neoplasms" for a volume that, for various reasons not relevant here, has yet to see the light. Before the manuscript could be prepared for the printer, the review was already somewhat out of date. This fact indicates that in the past few years the field of cytology has been a rapidly growing one. With constant improvements in techniques (Hungerford, 1955; Levan, 1956a; Ford and Hamerton, 1956b) older data are subjected to critical evaluation and much hitherto accepted information is now considered erroneous. The picture of the variation of chromosome numbers in somatic cells has consequently gained increasing clarity.

It is probably appropriate in this type of symposium to present a résumé of recent progress in the area so as to stimulate the formulation of working hypotheses, speculations, and to discuss the possibilities for future research. Most of the citations in this article will be limited to the more recent papers.

The most important question of all is what Beatty wrote for the title of his review (1954): "How many chromosomes in mammalian somatic cells?" Beatty was strongly in favor of the concept of chromosome inconstancy of mammalian somatic cells. However, most of the findings published before the last few years are in serious doubt. For instance, the series of papers by Timonen and Therman (1950, 1951) are now regarded as completely erroneous

[1] Section of Cytology, The University of Texas M. D. Anderson Hospital and Tumor Institute, Houston. These investigations were supported in part by The American Cancer Society Grant CP-84 and The Damon Runyon Memorial Fund Grant DRG-269.

by contemporary cytologists. Our data on chromosome variation of tissue culture cells, even though showing much narrower distributional spectra than theirs, are still highly exaggerated by observational mistakes.

There have been two main difficulties in determining the correct chromosome number of mammalian cells: (1) many cells may be broken due to the pressure in squash preparations. In Feulgen slides this difficulty is intensified by the colorless cytoplasm so that the intactness of the cell cannot be ascertained. This probably can also account for the fallacious data of Timonen. (2) The long chromosomes, especially the metacentric ones, often are difficult to count. For instance, the diploid chromosome number of man has long been recognized as 48. Cytologists seeing other numbers would question their own observations. Even with hypotonic solution treatment, which greatly increases the spreading of the chromosomes, my observation on this subject was still incorrect. In 1956 Tjio and Levan and Ford and Hamerton almost simultaneously declared that the diploid chromosome number of man is 46. This revised number has gained popularity in less than a year's time because the photomicrographs of Tjio and Levan are so clear that they leave no room for doubt. Many workers, including myself, after re-examining the human cells, agree with the aforementioned authors. Kodani's findings (1957) in which he found human individuals to have 46, 47, or 48 chromosomes, however, seems to cast a new cloud over this subject. It appears that the subject of normal chromosome number of man needs much more information before the arguments can be considered settled.

With the uncertainty of most of the reports dealing with chromosome numbers, discussion on the problem of their variation is impeded by the paucity of material. Newcomer and Brant's (1954) revolutionary report that the domestic fowl has only six pairs of true chromosomes, while the rest are heterochromatic bodies that may fuse or break up, may have rendered obsolete all the voluminous literature on the chromosomes of birds and reptiles. This leaves only a small fraction of published data on mammalian species during recent years that is usable. One generalization seems to be reasonable: somatic chromosome number does not vary greatly, except for true polyploidy. But aneuploids definitely exist, even though the fraction occupies only a very small proportion of the cell population in the body. The degree of aneuploidy is also exceedingly narrow. Recent personal communication with A. Levan, J. Schultz, G. Yerganian, and a number of others has confirmed this conclusion.

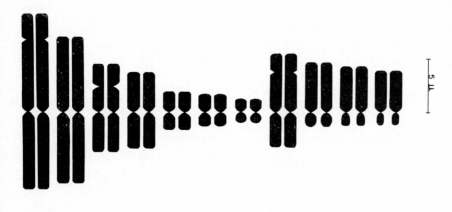

I   II   III   IV   V   VI   VII   VIII   IX   X   XI

Fig. 12. Idiogram of the diploid chromosomes of the Chinese hamster (colchicine pretreatment). (*Courtesy,* G. Yerganian.)

It is of special interest to note from the work of Yerganian (1952) that of the eleven pairs of chromosomes of the Chinese hamster (*Cricetulus griseus*) five are individually recognizable and the remaining six fall into two distinguishable groups of three; within these two groups the individual chromosomes can be differentiated with less certainty (Fig. 12). It is possible, therefore, to study the chromosomal constitution of cells with strictly diploid number. Studies of Yerganian and his associates on somatic tissues and tumors of this animal reveal that some of the cells were found to be quasidiploid, i.e., a cell with diploid number but not two complete chromosome sets. For instance, one chromosome may be represented three times and another only once, resulting in a $2n + 1 - 1$ condition. The discovery of quasidiploids, a special type of aneuploid, presents a problem to karyological workers on other species where such distinction cannot be made, because this type of cell may conceivably exist in any tissue of any species and yet escape the attention of cytologists. The presence of quasidiploids may raise the true proportion of aneuploids to diploids in normal tissues.

Heteroploids with highly unbalanced chromosomal constitutions have not been critically recorded in normal tissues. In regenerating liver of the rat, however, Makino and Tanaka (1953) and Tanaka

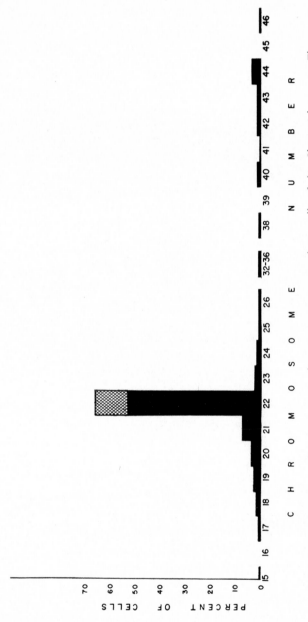

Fig. 13. Distribution of chromosome numbers in regenerating liver cells of the Chinese hamster. The cross-hatched area represents the quasidiploid fraction. (Courtesy, G. Yerganian.)

(1953) reported a relatively high proportion of cells containing chromosomes in the triploid range. They were probably heteroploids, not true triploids. Since it would be highly improbable for reputable cytologists like these to confuse diploids or tetraploids with triploids, there is little doubt that they did find such elements. In similar experiments Tonomura and Yerganian (1956) found few such cells in the regenerating liver of the Chinese hamster (Fig. 13) but a number of aneuploids. The question is whether these cells were formed after the regenerative process had started. In other words, during the period of regeneration a tremendous rate of cell division is generally observed and mitotic abnormalities are not uncommon. Some of these, such as multipolar spindles of polyploids, may give rise to heteroploids. Yet to attribute the production of all the heteroploids to the period of regeneration would seem to be overdogmatic. Even at rather early stages of regeneration, say 20 hours, Makino and Tanaka observed such "triploids." It could only be reasonable to conclude that some of these cells existed before the surgical operation. Cytologists studying chromosomes are all handicapped by the fact that they can only study the cells in mitotic division. Should cells of different chromosomal constitution show a different rate of mitosis, the chromosomal spectrum, even estimated from large and most unbiased samples, would not absolutely reflect the true picture of the tissue. It is possible that heteroploids under normal conditions do not enter mitosis as often as the diploids, but metabolically and functionally they may still be active. Only under adverse conditions, such as in the liver after partial extirpation, some stimulant may induce them to divide. In neoplasms we often speak of the *stemline cells;* this term denotes the elements that give rise to the cell lineage of the tumor. We shall discuss this point more fully later. But here can we not say that in normal tissues the diploids are the stemline cells, which would also mean that odd types of cells are not deprived of their right of existence?

The Chromosomes of Neoplasms

When we take a glance at the literature on the karyology of neoplasms, we can immediately notice that it would be difficult to search for a neoplasm that did not contain chromosomal anomalies. The lone report that claims to have found only diploid chromosome number is the one by Ohno *et al.* (1957) on a mouse lymphoma.

However, this was verbally corrected by the authors and the tumor actually is characterized by a count of 41, $2n + 1$. It is not at all impossible that a tumor can be found to show no demonstrable morphological deviations such as numerical variation or gross structural changes and yet have accumulated numerous minor alterations not detectable by ordinary cytological methods (Tjio and Levan, 1956b; Makino and Sasaki, 1958). From present technical knowledge it seems that the subdiploid rectal carcinoma reported by Koller was probably also a mistake. The revision of the diploid chromosome number of man almost demands a complete discard of old literature, as far as those near diploid tumors are concerned. From now on the chromosome counts on such matters must be unequivocal or they will be worse than useless.

The bulk of data on the chromosomes of neoplasms comes from the studies of mouse and rat tumors. It is probably incidental that a number of rat tumors show a reduction of diploid chromosome number coupled with numerous structural changes, whereas in most mouse tumors an increase of chromosome number has been noted. Some of the mouse neoplasms are near diploid, some distinctly hyperdiploid, some subtetraploid, some tetraploid, and some hypertetraploid. This is, of course, to describe each tumor expediently by its most representative group. Actually, in practically every population there are variants with other chromosome numbers. Indeed, tumors such as the Ehrlich carcinoma of the mouse have produced many sublines, each possessing a modal chromosome number of its own.

Employing various methods of abusing the cells, such as cold storage, starvation, small dose inoculum, Kaziwara (1954) was able to transform the hyperdiploid Ehrlich line from Heidelberg into many near tetraploid strains. In Fig. 14 the strains that were produced by various methods all changed to polyploid but a small fraction of cells remained near diploid. The situation was the reverse of that in the original population, which contained mainly hyperdiploid with a small fraction of polyploid cells (rightmost column in Fig. 14). This demonstrates that the cell population, even though at one time dominated by one type of cell, is able to readjust itself when adverse conditions arise. In other words, the various genotypes of the population may serve as material for emergency measures. After the treated strains were carried in the regular manner some of the lines gradually reverted to the original constitution.

In the light of recent information the concept of stemline needs to be modified slightly. The stemline cell contains the type of genotype that best fits the given environment. These cells may be the most vigorous and most proliferative. However, other genotypes,

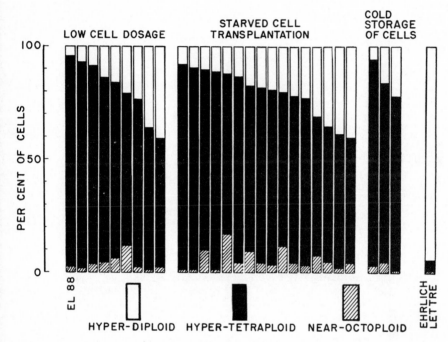

Fig. 14. Comparison of percentage frequency of hyperdiploid, hypertetraploid, and near-octoploid cells among 28 polyploid subline derivatives from the original hyperdiploid Heidelberg Ehrlich ascites tumor, shown on the extreme right. (After Kaziwara, 1954.)

serving as reserves, may take over the population when the present stemline cells are unfit for the changed environment. By that time the genotype that dominates will become the stemline. Genetically, ascites tumors or tissue culture bottles simulate the situation of selection and adaptation of genotypes in nature.

It is reasonable to assume then that the cell populations of tumors are ever-changing. They may seemingly be stabilized at one period but not fixed. The proportion of the existing cell types may change; the environment may change; and new mutations and chromosomal aberrations may come into the scene. Levan (1956b) witnessed the gradual shift of average chromosome number per cell of both the Ehrlich and the Krebs-2 carcinomata within 200 transplantation generations. When the Ehrlich cancer was transplanted into hamsters, the stemline shifted from 80 to 76 (Ising, 1955). However, when this line returned to the mouse, this new number persisted.

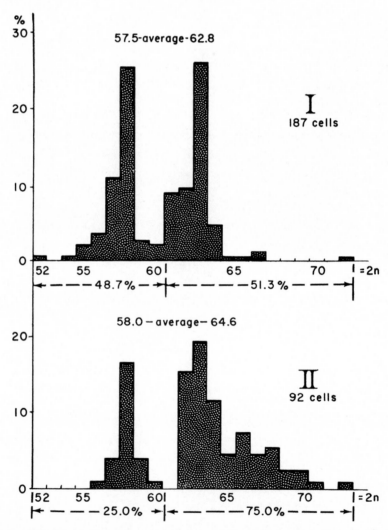

Fig. 15. Distribution of chromosome numbers of two samples of a human ascitic cystocarcinoma of the ovary. (After Hansen-Melander et al., 1956.)

The cytology of human cancers is not drastically different from that of experimental animals. Chromosomal changes have repeatedly been noted either in number or in structure or both (Ising and Levan, 1957; Levan, 1956a). The cell populations may also show shifts of chromosome numbers. An example is the report of Hansen-Melander *et al.* (1956), who observed such a shift in a case of a human ascitic cystocarcinoma of the ovary in a relatively short period (Fig. 15). Their second sample emphatically suggests the trend of population adjustment toward the higher numbered classes. It is obvious from Fig. 15 that although the cell populations were most characteristically represented by two modal numbers, 58 and 63, other genotypes were still present in a relatively large quantity. In other words, the populations contained, if each number represented a genotype, 15 or 16 genetically different potentialities. It is obvious that each number could represent a multitude of combinations of various chromosomes. Some of our unpublished results with marker chromosomes give evidence in this direction.

The Chromosomes of Cells *in Vitro*

An interesting aspect of the problem of tumor chromosomes is the study of cells *in vitro*. After the wide use of the famous HeLa strain of human cervical carcinoma, which indicated the feasibility of growing human cells in a perpetual fashion, human cell strains mushroomed. The Tissue Culture Association even set a committee to deal with the collection, cataloguing, and nomenclature of the cell lines. Hsu's description of HeLa chromosomes, confirmed by Petrakis by his microspectrophotometry data, has established HeLa as a subtetraploid (cf. Hsu and Moorhead, 1957). However, as revealed by their histograms, practically any heteroploid number is possible among the HeLa cells. This adequately points out that the cell line is genetically a mixture. Indeed, by studying Puck's clonal derivatives of HeLa, Chu and Giles (1958) have found that a stemline condition exists in most clones studied. These clones can be distinguished from HeLa and from each other by either chromosome number or chromosome morphology. A more interesting phenomenon is that most of the clone strains of HeLa showed a high proportion of the modal number, whereas one clone displayed a dispersed spectrum as wide as the mother HeLa. Perhaps one genotype may be more stable in producing mitotic abnormalities than another, a phenomenon worth further investigation.

Not only are all the human tumor cell strains so far studied heteroploid, but the strains of nonmalignant origin show the same complexity of chromosomal patterns (Hsu and Moorhead, 1957; Ber-

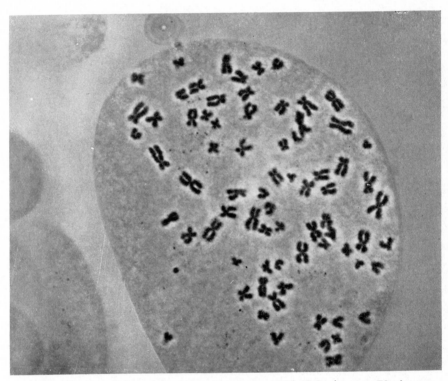

Plate III. A cell from human amnion strain A185-11C, showing 73 chromosomes. Colchicine and hypotonic solution pretreated, aceto-orcein squash, dark-phase microscopy.

man *et al.*, 1957). Plate III represents a cell from an amnion strain A185–11C (Dunnebacke and Zitcer) and Plate IV shows a cell of strain L (mouse) with a long, dicentric element. To the existing lists we have recently added the following lines of normal origins:

> Amnion strain A185–11C (supplied by Wendell Stanley and Thelma D. Dixon)
> Amnion strain A205–14C (supplied by W. Stanley and T. D. Dixon)
> Human skin strain NCTC (clone) 2414–12–4 (supplied by W. R. Earle)
> Human heart strain (supplied by Anthony J. Girardi)

All these strains are again heteroploid.

It is well known that normal tissues are principally made of diploid cells. Even the primary cultures show excessive numbers of diploids. The fact that the established cell strains derived from normal tissues are completely heteroploid indicates a drastic transformation in the history of each strain. Hsu *et al.* (1957) fol-

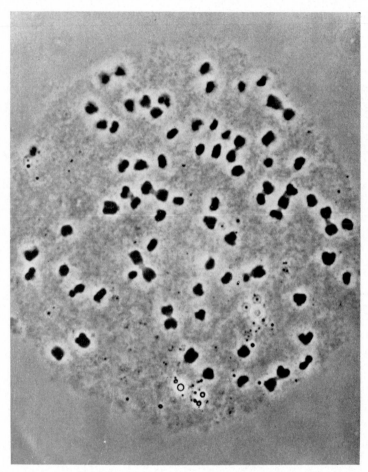

Plate IV. A cell from mouse strain L, showing 67 chromosomes with a dicentric element. Colchicine and hypotonic solution pretreated, aceto-orcein squash, dark-phase microscopy.

lowed cytologically strain Mayes from its original tissue (a normal human synovial knee lining) through each subculture until a total heteroploid population was reached. At subculture 5 the dividing cells were still typical diploids with 46 chromosomes. At subcultures 8 and 9 such cells completely disappeared. Instead, dividing cells were secondary heteroploids, showing more than a hundred chromosomes. Unfortunately, the preparations for subcultures 6 and 7 contained no mitotic figures to furnish information as to whether the transformation was a sudden or a gradual replacement.

To interpret the phenomenon of heteroploid transformation is difficult without making speculations. The fact that cell lines finally turn to heteroploid seems to suggest that heteroploid cells, at least with a certain combination of genetic material, are more adaptable to the tissue culture conditions than the euploids. To cite from Hsu and Moorhead (1957):

Tissue-culture media and conditions must be regarded as abnormal in a strict sense. Even though they can support migration and mitosis of cells, in vitro conditions fail in certain ways to duplicate the in vivo environment. Cellular relationships, for instance, obviously deviate from normal. Many attempts to establish cell strains have failed, even with embryonic material. Probably the normal cells (presumably diploid) cannot survive and proliferate for long periods of time in vitro, and the cell population gradually loses its reproductive ability, whereas heteroploids, with their special combinations of chromosomes and therefore special genotypes, may adapt to the new environment. It is of interest to cite the work on rats by Makino and Tanaka, which showed that regenerating cells in situ contained a relatively high frequency of heteroploids after partial extirpation of liver. When tissues are set up in culture vessels, trauma is unavoidable. This may stimulate some production of abnormal mitoses and consequently abnormal chromosome numbers. During the early periods of cultivation, abnormal mitoses may repeatedly take place, thus increasing the frequency of heteroploids. As the vigor of the diploids finally subsides, the stage of "sluggish growth" is reached. This period is probably very important in the history of a strain, for it may represent the stage of transformation. Heteroploids, which are fewer in number, now begin to dominate the population and replace the diploids. Vigorous growth is apparent after this transition era due to the tremendous rate of mitosis in the new type of cells. In other words, the phenomenon of strain establishment is likely to be a result of selection within a cell population. Cultures failing to develop into cell strains may contain no heteroploids or may contain heteroploids with a genotype that does not possess an adaptational advantage over the diploids. It is possible also that the culture media used might be one of the factors influencing the spectrum of heteroploidy.

The reason for the adaptational advantage of heteroploids is not known. From what we learned from classic genetic concepts, these cells should not have existed. Yet they not only exist, but thrive, and possess superior metabolic abilities. Perhaps the partial anaerobiotic condition suggested by Goldblatt and Cameron (1953) plays a significant role in inhibiting the diploids and in stimulating the heteroploids. These authors employed a special apparatus to control the gaseous environment of the roller tube culture. In one set of experiments the atmosphere was regular air, while in another, daily shock with pure nitrogen was performed to create an intermittent anaerobiosis on the rat cells. The exposed cultures later converted into neoplasms but the controls did not. The contemporary bottle cultures may provide conditions for a partial anaerobiosis, which may be responsible for the selection of genotypes. Suppose from anomalous divisions a number of heteroploid cells arise. Naturally they differ from each other in their genome. Some may contain certain chromosomes in surplus and some may contain other chromosomes in surplus. Reasonably speaking, not all the cells should be metabolically superior over the diploids. Some of them may even be inferior to the diploids and hence be eliminated from the population. Now if the anaerobic condition is one of the major factors responsible for selection of genotypes, those cells with extra chromosomes that happen to contain genes for performing anaerobic glycolysis will be better able to grow and will gradually be selected to take over the cell population. If this is true, the transformed heteroploid cell lines should show higher activity of anaerobic glycolysis than their normal counterparts. It is indeed of interest to mention the work of A. L. Bresson (personal communication), who studied the ability of cell strains in performing anaerobic glycolysis and found that strains like HeLa, L, McCoy, etc., all showed a higher fermentation rate than normal tissues.

Another phenomenon closely related to the heteroploid transformation *in vitro* is the aforementioned conversion to malignancy. Moore *et al.* (1956) and Southam *et al.* (1957) found that not only the neoplastic strains, such as HeLa, the HEP's, etc., but also the nonmalignant cell lines, such as Ch-L, Ch-C, etc., produced "tumors" upon inoculation into cancer patients at terminal stages. The relationship between malignancy and chromosomal changes seems to be brought one step closer, viz., when normal tissues change their chromosomal constitution, the tissues become malignant. Of course there are still criticisms against this oversimplified conclusion that all the cytologically transformed cell strains are neoplasms. There are also criticisms attacking the validity of regarding the "tumors" produced in patients as true "cancers." Decisive proof remains to be obtained; but from what we can infer from the results on malignant conversion of animal tissue cultures and from pathological examination on the "tumors" produced in the Sloan-Kettering Institute, the artifically induced growth under the skins of patients were most likely to be neoplasms. At any rate, we can safely make one conclusion, i.e., all the "normal" cell strains studied are abnormal, whether they are truly neoplasms or not. It is not unreasonable to infer that with many extra chromosomes (consequently genic products), the metabolism will be drastically changed.

Outlook of the Problems

As has been discussed before, most data on chromosomal variation are not reliable. Therefore, even fundamental problems such as the proportion of aneuploidy and heteroploidy among various normal tissues, the chromosomal features of human and animal neoplasms *in vivo*, heteroploid transformation *in vitro*, and a number of related problems are still lacking critical supporting data. Also, conclusive proof is urgently needed to demonstrate that the cytologically transformed strains are genuine cancers. If they are, which of the two phenomena, transformation to heteroploidy or transformation to neoplasm, occurs first? Does one cause the other or is there any causative relation at all between them? In other words, it is of cardinal value to examine critically the concept that certain types of unbalance of the genetic make-up (chromosomes) of a cell causes the cell to become neoplastic. This goes back to the old Boveri hypothesis, which once again in recent years has become one of the leading theories with a number of adherents.

To the workers in the field of mammalian cytology it is perhaps more glamorous, more impressive, and hence more satisfying to investigate human cells, human chromosomes, and human cancers. As a matter of fact, researchers are more handicapped by working with this material because (1) it is difficult to obtain as constant a supply of human tissues as experiments call for, (2) it is extremely difficult to obtain legally permitted human volunteers for purposes of experimentation, especially for investigators not working in hospitals, (3) even if volunteers can be arranged, their use is rather limited, and (4) karyologically human chromosomes are unfavorable material. The diploid number is high; and the chromosomes are all metacentric, giving little criterion for recognizing individual elements, except the difference in length. Even mice and rats are not ideal objects in this respect. The best available material to our knowledge is the Chinese hamster. With 11 pairs of morphologically distinguishable chromosomes and with the ease in breeding worked out by Yerganian, it is anticipated that this animal will sooner or later dominate a number of experimental animal quarters in research institutes.

For tissue culture work the Chinese hamster could offer endless possibilities for future research exploration. If heteroploid transformation can also be obtained (there is no reason that it would not happen), it would then be possible to determine which chromosome or chromosomes are duplicated. By employing the methods of Puck and his associates to isolate clone strains (Puck and Marcus, 1955), cell lines of relatively homogeneous genetical background can be carried and many experiments can be made. In fact, Puck and Fisher (1956) have reported inherent nutritional differences between two clone strains of HeLa, S1 and S3.

It is about time for us to correlate cytological findings with biochemical data to interpret the role of certain chromosomes in metabolism. Selection of mutants in cell strains (e.g., Swim and Parker, 1957; Haff and Swim, 1957) can also be checked cytologically if the material used is cytologically favorable. Preliminary data of Kit (personal communication) on the differences in amino acid content among strains of the Novikoff hepatoma which differ cytologically well illustrate such a beginning.

References

BEATTY, R. A. 1954. How many chromosomes in mammalian somatic cells? *Internat. Rev. Cytol.* 3: 177–197.

BERMAN, L., C. S. STULBERG, and F. H. RUDDLE. 1957. Human cell culture. Morphology of Detroit strains. *Cancer Research* 17: 668–676.

CHU, E. H. Y., and N. H. GILES. 1958. Comparative chromosomal studies on mammalian cells in culture. I. The HeLa strain and its mutant clonal derivatives. *J. Nat. Cancer Inst.* 20: 383–401.

FORD, C. E., and J. L. HAMERTON. 1956a. The chromosomes of man. *Nature 178:* 1020–1023.

FORD, C. E., and J. L. HAMERTON. 1956b. A colchicine, hypotonic citrate, squash sequence for mammalian chromosomes. *Stain Technol. 31:* 247–251.

GOLDBLATT, H., and G. CAMERON. 1953. Induced malignancy in cells from rat myocardium subjected to intermittent anaerobiosis during long propagation in vitro. *J. Exp. Med.* 97: 525–552.

HAFF, R. F., and H. E. SWIM. 1957. Isolation of a nutritional variant from a culture of rabbit fibroblasts. *Science 125:* 1294.

HANSEN-MELANDER, E., S. KULLANDER, and Y. MELANDER. 1956. Chromosome analysis of a human ovarian cystocarcinoma in the ascites form. *J. Nat. Cancer Inst. 16:* 1067–1081.

Hsu, T. C., and P. S. Moorhead. 1957. Mammalian chromosomes in vitro. VII. Heteroploidy in human cell strains. *J. Nat. Cancer Inst. 18:* 463–471.

Hsu, T. C., C. M. Pomerat, and P. S. Moorhead. 1957. Mammalian chromosomes in vitro. VIII. Heteroploid transformation in the human cell strain Mayes. *J. Nat. Cancer Inst. 19:* 867–873.

Hungerford, D. A. 1955. Chromosome numbers of ten-day fetal mouse cells. *J. Morphol. 97:* 497–510.

Ising, U. 1955. Chromosome studies in Ehrlich mouse ascites cancer after heterologous transplantation through hamsters. *Brit. J. Cancer 9:* 592–599.

Ising, U., and A. Levan. 1957. The chromosomes of two highly malignant human tumors. *Acta Pathol. et Microbiol. Scand. 40:* 13–24.

Kaziwara, K. 1954. Derivation of stable polyploid sublines from a hyperdiploid Ehrlich ascites carcinoma. *Cancer Research 14:* 795–801.

Kodani, M. 1957. Three diploid chromosome numbers of man. *Proc. Nat. Acad. Sci. 43:* 285–292.

Koller, P. C. 1947. Abnormal mitosis in tumors. *Brit. J. Cancer 1:* 38–47.

Levan, A. 1956a. Chromosome studies on some human tumors and tissues of normal origin, grown in vivo and in vitro at the Sloan-Kettering Institute. *Cancer 9:* 648–663.

Levan, A. 1956b. Chromosomes in cancer tissue. *Ann. N. Y. Acad. Sci. 63:* 774–792.

Makino, S., and M. Sasaki. 1958. Cytological studies of tumors. XXI. A comparative idiogram study of the Yoshida sarcoma and its subline derivatives. *J. Nat. Cancer Inst. 20:* 465–488.

Makino, S., and T. Tanaka. 1953. Chromosome features in the regenerating rat liver following partial extirpation. *Tex. Rep. Biol. Med. 11:* 588–592.

Moore, A. G., C. M. Southam, and S. S. Sternberg. 1956. Neoplastic changes developing in epithelial cell lines from normal persons. *Science 124:* 127–129.

Newcomer, E. H., and W. A. Brant. 1954. Spermatogenesis in the domestic fowl. *J. Hered. 45:* 79–87.

Ohno, S., R. T. Jordon, and R. Kinosita. 1957. Chromosomes of lymphocytic neoplasm of strain AKR mice. *Proc. Am. A. Cancer Res. 2:* 236.

Puck, T. T., and P. I. Marcus. 1955. A rapid method for viable cell titration and clone production with HeLa cells in tissue culture: The use of X-irradiated cells to supply the conditioning factors. *Proc. Nat. Acad. Sci. 41:* 432–437.

Puck, T. T., and H. W. Fisher. 1956. Genetics of somatic mammalian cells. I. Demonstration of the existence of mutations with different growth requirements in a human cancer cell strain (HeLa). *J. Exp. Med. 104:* 427–434.

Southam, C. M., A. E. Moore, and C. P. Rhoads. 1956. Homotransplantation of human cell lines. *Science 125:* 158–160.

Swim, H. E., and R. F. Parker. 1957. Isolation of nutritional variants from a mammalian cell culture. *Fed. Proc. 19:* 435.

Tanaka, T. 1953. A study of the somatic chromosomes of rats. *Cytologia 18:* 343–355.

Therman, E., and S. Timonen. 1951. Inconstancy of the human somatic chromosome complement. *Hereditas 37:* 266–279.

Timonen, S. 1950. Mitosis in normal endometrium and genital cancer. *Acta Obst. and Gynecol. Scand. 31* (Suppl. 2): 1–50.

Timonen, S., and E. Therman. 1950. Variation of the somatic chromosome number in man. *Nature 166:* 995–996.

Tjio, J. H., and A. Levan. 1956a. The chromosome number of man. *Hereditas 42:* 1–6.

Tjio, J. H., and A. Levan. 1956b. Comparative idiogram analysis of the rat and the Yoshida rat sarcoma. *Hereditas 42:* 218–234.

Tonomura, A., and G. Yerganian. 1956. Aneuploidy in the regenerating liver of the Chinese hamster. *Genetics 41:* 664–665.

Yerganian, G. 1952. Cytogenetic possibilities with the Chinese hamster, *Cricetulus Barabensis griseus*. *Genetics 37:* 638.

4

Nuclear and Cytoplasmic Changes in Tumors

GEORGE KLEIN AND EVA KLEIN [1]

It would be clearly impossible within the limits of this symposium paper to give a complete critical survey of the various nuclear and cytoplasmic changes that have been described in different neoplasms. The authors have therefore chosen to discuss a few selected topics only. Attention is mainly focused on some recent studies on the isoantigenic composition and the chromosome cytology of tumor cells.

Immunogenetic Markers

The nuclear constitution of neoplastic cells can be approached either by genetic or by cytological methods. As with somatic cells in general, the genetic methods are indirect and they are at present limited to a group of genes concerned with isoantigenic specificity, usually called histocompatibility genes. Even here, only the isoantigenic products of the genes can be studied and not the genes themselves. The information obtained may permit some conjecture about the presence and behavior of the genes, at least in some carefully selected cases. Indirect as it is, such information nevertheless may be of interest for studies on the cytogenetics of tumor cells and possibly also with regard to the cytogenetic changes that may occur during the process of carcinogenesis. We shall try to demonstrate this with some specific examples.

[1] Department of Tumor Biology and Department of Cell Research and Genetics, Karolinska Institutet, Stockholm, Sweden. The experimental work quoted in this paper has been supported by grants from the Swedish Cancer Society and by Grant C-3700 from The National Cancer Institute, U. S. Public Health Service.

Several decades ago, it was shown by Strong and Little (1920), Little and Strong (1924), Strong (1929), Cloudman (1932), and others that different primary tumors of closely similar histological origin and arising in the same inbred strain of mice (or, in certain cases, in the same individual) may show differences with regard to their histocompatibility characteristics. With the advent of isogenic resistant lines of mice, developed by Snell (1948), the elaboration of suitable serological methods for the study of the iso-antigenic components by Gorer and his collaborators (cf. review of Gorer, 1956), and an increased understanding of the immunogenetic basis of tissue and tumor transplantation in general, it is now possible to create situations in which tumors can be studied with regard to the presence or absence of the products of *single* histocompatibility genes. We have carried out such a study recently (Klein *et al.*, 1957), the results of which will be described briefly in the following.

The isogenic resistant lines of Snell (1948, 1955) are inbred strains of mice genetically identical with the exception of a single gene locus. The single gene differentiating the different lines is a histocompatibility gene that causes one line to resist transplants from the other, due to a homograft reaction directed against the foreign isoantigens, the specificity of which is determined by the foreign allele at the histocompatibility locus in question.

We have used four isogenic resistant lines having a strain A/Sn genetic background but differing at the histocompatibility-2 (H-2) locus. H-2 is perhaps the strongest and certainly the best-known histocompatibility gene of the mouse (Snell *et al.*, 1953). Our procedure has been to cross two of the isogenic resistant lines, A/Sn (genotype: $H\text{-}2^A H\text{-}2^A$) and A.SW (genotype: $H\text{-}2^S H\text{-}2^S$) and induce fibrosarcomas in the F_1 hybrids by the intramuscular injection of methylcholanthrene. Since both parental lines have the same isogenic background, such hybrids are homozygous, at least theoretically, with respect to all gene loci except H-2, in which the parental strains differ, making the hybrids heterozygous. As a rule, neoplasms that originate in hybrids of this type do not grow progressively in either of the parental strains, since each parental strain is capable of reacting against the isoantigenic products of the foreign H-2 allele derived from the other parental strain and both alleles are present in the tumor cells. Such tumors can be maintained only by transplantation to F_1 hybrids having the same genetic constitution as the animal in which they arose. However, if a specific mutation occurs at one of the two H-2 loci in a tumor cell that proceeds in the direction of the noncorresponding parental type (or, alternatively, one that renders the gene isoantigenically inactive for the non-corresponding parental strain), it will confer an absolute selective advantage upon the bearer whenever the cell population is tested in that particular parental strain, provided that the homograft re-action destroying the unchanged and incompatible cells does not also destroy the mutant in a nonspecific way.

We have carried out some model experiments (Klein and Klein, 1956) by mixing tumor cells of known genotypes in various pro-portions and inoculating the mixtures into appropriate hosts. As a rule the inocula consisted of a large number of incompatible cells containing a very small number of randomly admixed compatible cells. The compatible cells were genetically identical with the host, while the incompatible ones were rejected because of a single gene difference between the hosts and the tumor cells with regard to the allelic substitution at one of the two H-2 loci. These model experi-ments have confirmed that the system indeed possessed the expected unique qualities of specificity and sensitivity. The homograft re-action that destroyed the large population of incompatible cells did not succeed in destroying or damaging the small compatible frac-tion: on the contrary, their growth was being stimulated by some factor involved in the concomitant homograft reaction.

The model experiments having given this encouraging result, we tested five induced fibrosarcomas, all of which had been induced by methylcholanthrene in A $\times$ A.SW F$_1$ males. They showed highly uniform histological characteristics. All tests were carried out on the primary tumors themselves or on tumors carried for less than five transfer generations in mice of the original F$_1$ genotype. Besides inoculations into the genotype of origin, all tumors were tested in the two parental strains, A and A.SW, and also in some other iso-genic resistant mice, carrying foreign H-2 alleles. These were the A.BY (genotype: H-2^{B}H-2^B) and the A.CA (genotype: H-2^{F}H-2^F) lines. A few tests were also made in A $\times$ A.BY and A $\times$ A.CA F$_1$ hybrids.

Three tumors, called DSWB, MNSB, and MSWG, were found to behave in strict accordance with the genetic theory of transplanta-tion; they grew only in A $\times$ A.SW F$_1$ hybrids but regressed in both parental strains (Table 1). This was also true if the inoculum was irradiated *in vitro* prior to testing with a dose of between 300 and 500 r X-rays. One tumor, MSWC, grew rather indiscriminately (although never in 100 per cent of cases) in both parental strains and also in other isogenic resistant mice carrying foreign H-2 alleles.

TABLE 1

Tests on Sarcomas Induced by Methylcholanthrene in A × A.SW F₁ Mice

Tumor	Progressive Growth in					
	A×A.SW F₁	A	A.SW	A.BY	A.CA	A×A.BY
D S W B	51/51	0/8 0/15*	0/15 0/20*			0/2
M N S B	20/20	0/17 0/5*	0/8 0/6*	0/5		
M S W G	8/8	0/12	0/15			
M S W C	75/75	29/35 3/8* 4/12**	61/78 4/11* 3/8**	10/16	6/9	5/7

° After irradiation of inoculum *in vitro* (300–500 r).
°° After irradiation followed by one passage in A.SW mice.

Progressively growing MSWC tumors that appeared after inoculation into the A.SW parental strain still grew upon subsequent testing in both the A and the A.SW strain.

The most interesting behavior was shown by the fifth sarcoma, called MSWB (Table 2, first column). This tumor grew progressively in 100 per cent of the F₁ hybrids of the type of origin and in 0 per cent of the parent strain A, as expected. However, it was found that it also grew progressively in 11 of 63 animals of the A.SW parent strain, a result that was unexpected. The percentage of takes in backcross hybrids was close to the theoretical 50 per cent. As a rule, the tumor did not grow in animals carrying the unrelated H-2ᴮ and H-2ᶠ alleles, with the exception of 2 of 21 A × A.BY F₁ hybrids. One of the latter two tumors was tested again in A × A.BY F₁ mice and in other genotypes and behaved quite similarly to the original tumor; it grew in (A × A.SW) F₁ mice but not in other genotypes.

TABLE 2

Growth of Original MSWB Tumor Line, Its Irradiated and Its Selected Sublines
After Inoculation into Mice of Various Isogenic Resistant Lines and Their Hybrids,
Differing Theoretically at the H-2 Locus Only

M S W B   s a r c o m a

origin A x A.SW F₁

genotype: H-2^A H-2^S

	control	irradiated	selected
A (H-2^{A}H-2^A)	0/44 (0%)	0/29 (0%)	0/39 (0%)
A.S W (H-2^{S}H-2^S)	11/63 (17%)	24/49 (49%)	133/146* (91%)
AxA.SW F₁ (H-2^{A}H-2^S)	47/47 (100%)	28/28 (100%)	52/52 (100%)
(AxA.SW) xA.SW BC	26/47 (55%)		69/74 (93%)
A.BY (H-2^{B}H-2^B)	0/32 (0%)	0/8 (0%)	0/27 (0%)
AxA.BY (H-2^{A}H-2^B)	2/21	1/3	1/15
A.SW x A.BY F₁ (H-2^{S}H-2^B)	0/7		
A CA (H-2^{F}H-2^F)	0/7	0/3	1/27
AxA.CA F₁ (H-2^{A}H-2^F)	0/3	0/3	0/4

* The selected tumor grew progressively also in 41 out of 49 additional A.SW
mice that have been immunized against the control line.

Concurrently with some of these experiments, part of each cell
suspension used for inoculation was irradiated *in vitro* with 300 or
500 r and subsequently was tested in a similar manner (Table 2,
second column). The percentage of takes in A.SW mice rose to 24
of 49, i.e., from 17 to 49 per cent; the difference is highly significant.
No takes were obtained in the A-strain.

A number of the tumors that appeared in A.SW mice inoculated either with the irradiated or the control tumor, were selected and tested further in A.SW mice and other genotypes. Since there was no systematic difference between tumors from the irradiated or control lines and several tests gave essentially the same result, the results of these experiments are pooled in the third column of Table 2. For clarity, all tumors that grew in A.SW mice and that were subsequently tested or carried further will be called variants. The variants grew in 91 per cent (133 of 146) of A.SW mice, a highly significant increase, when compared with either the control or the irradiated tumor. The percentage of takes in backcross mice increased to 93 per cent. The variant tumors were consistent in their failure to grow in the other parental strain, A, or in genotypes carrying foreign alleles of H-2, thereby demonstrating that they did not become "nonspecific."

The behavior of the variant tumor might be explained by assuming that the $H-2^A$ allele was lost from the tumor cells or that it mutated to $H-2^S$, or to some third, indifferent, form. This is not the only possible mechanism, however. It would be conceivable that the $H-2^A$ allele is still present in the nuclei of the variant cells but that they have changed with regard to their isoantigenicity, i.e., their capacity to provoke the formation of anti-$H-2^A$ antibodies in protective titer. Alternatively, their ability to resist such antibodies may have increased. Changes of this type have been described and could be unmasked by testing the tumor for growth in mice preimmunized against the isoantigenic factors in question (Amos *et al.*, 1955). The variant tumor was therefore tested in A.SW mice in which the original MSWB sarcoma had previously regressed or which had been challenged previously with tumors of strain A origin. These animals have thus been immunized against the specific isoantigenic products of $H-2^A$. The variant tumor was able to grow in 41 out of 49 animals, thus indicating that, at least in the majority of the cases, it no longer contained receptors sensitive to the action of anti-$H-2^A$ antibodies.

Another means of demonstrating the presence or absence of the specific isoantigens of the $H-2^A$ group in the variant cells was made possible by the serologic methods developed by Gorer and his associates (cf. review of Gorer, 1956). In one series of experiments, isogenic resistant A.BY or A.CA mice, carrying the foreign $H-2^B$ or $H-2^F$ alleles, respectively, were hyperimmunized against the original or the variant tumor. As shown in Tables 3 and 4, their sera were found to contain both hemagglutinins and cytotoxic antibodies against A.SW cells carrying the H-2^s allele. On the other hand, only mice immunized with the *original* tumor showed the presence of

TABLE 3

Effect of Isoimmune Serum from A.BY Mice on Red Cells (Hemagglutination) and Lymphocytes (Cytotoxic Effect) of A and A.SW Mice *

	A. BY Anti—MSWB Serum			
	Unabsorbed		Absorbed with A.SW Tissue	
	Hemagglutinin Titer	Cytotoxic Effect	Hemagglutinin Titer	Cytotoxic Effect
A-cells	≧512	95	512	91
A.SW-cells	≧256	96	0	0

	A. BY Antivariant Serum			
	Unabsorbed		Absorbed with A.SW Tissue	
	Hemagglutinin Titer	Cytotoxic Effect	Hemagglutinin Titer	Cytotoxic Effect
A-cells	≧512	46	0	0
A.SW-cells	≧128	95	0	0

* The cytotoxic effect has been expressed as per cent cells killed after an incubation of one hour as related to the control.

specific anti-H-2^A antibodies (i.e., antibodies that would react with A cells even after having been absorbed with A.SW tissue to remove components directed against the isoantigens common to A and A.SW cells), while such antibodies were regularly absent from the sera of mice immunized against the *variant* tumor.

In another series of experiments (Table 5) the original and the variant tumors were compared for their capacity to absorb specific anti-H-2^A antibodies from hyperimmune sera produced by immunizing A.SW mice with an ascites tumor of strain A origin (carcinoma TA3). Both hemagglutinins and cytotoxic antibodies were absorbed

TABLE 4

Effect of Isoimmune Serum from A.CA Mice on Red Cells and Lymphocytes of A and A.SW Mice

	A.CA Anti-MSWB Serum			
	Unabsorbed		Absorbed with A.SW Tissue	
	Hemagglutinin Titer	Cytotoxic Effect	Hemagglutinin Titer	Cytotoxic Effect
A-cells	$\geqq 512$	71	256	44
A.SW-cells	$\geqq 512$	92	0	10

	A.CA Antivariant Serum			
	Unabsorbed		Absorbed with A.SW Tissue	
	Hemagglutinin Titer	Cytotoxic Effect	Hemagglutinin Titer	Cytotoxic Effect
A-cells	$\geqq 512$	12	0	0
A.SW-cells	$\geqq 512$	94	0	2

TABLE 5

Comparison of Regular (Genotype: H-2^{A}H-2^S) and Variant (Presumed Genotype: H-2^{S}H-2^S) MSWB Tumor for Ability to Absorb Agglutinins and Cytotoxic Antibodies from Anti-H-2^A Immune Sera

Antiserum	Absorbed with	Effect on A-cells	
		Hemagglutinin Titer	Cytotoxic Effect
A.SW anti-A	—	$\geqq 512$	92
"	MSWB	0	7
"	Variant	256	84
A.SW anti-A	—	$\geqq 512$	77
"	MSWB	0	3
"	Variant	256	82

regularly by tumors of the original line, whereas the variant line failed to show any activity.

Taken together, these results suggest very strongly that the specific isoantigenic products of the H-2^A gene are absent from the variant cells. The most plausible explanation of these findings is loss or mutation of the H-2^A allele itself.

The variant tumor was recovered repeatedly from different positive A.SW mice inoculated with the original line, using different samples of inocula from the primary tumor itself or its first five transfer generations. The question arose whether we are dealing with independent mutations or with just one or a few variant clones, already present in the original tumor and being maintained at a low but more or less constant frequency. Such a clone or clones might be eliminated by using a small number of cells for inoculation. A series of experiments was therefore carried out in which from 40 to 50 cells were inoculated into A × A.SW F$_1$ hybrids and the resulting tumors were tested in A.SW mice. Such tumors still gave 10 to 15 per cent takes in A.SW mice and the majority of these takes turned out to be true variants upon further testing. It would thus appear that the change from the original to the variant cell type is a fairly frequent event and our results are not merely due to the repeated selection of the same pre-existent clone(s) but rather to the ability of the MSWB tumor to throw off the variant type continuously from its main stemline. It is puzzling that the change has always been found to proceed in the direction of the A.SW parent and in no case towards the A-parent; it would appear that this tumor has some particular cytogenetical feature that makes it especially liable to lose H-2^A but not H-2^S. Recent work indicates that the H-2 alleles are in fact pseudoalleles, since crossing over may occur within one H-2 group (cf. Gorer, 1956). This would suggest that the change in the MSWB tumor is not due to point mutation. Loss of a whole chromosome or a chromosome segment would be a more likely explanation.

Taken as a whole, the experiments on the five sarcomas, induced by the same carcinogen in the same tissue of the same genotype, indicate that such tumors may differ with regard to the behavior of their H-2 system as early as at the primary stage. Our findings are comparable to those of Sachs and Gallily (1956). These authors have tested 12 primary carcinomas and sarcomas. Eight tumors were strain specific and did not give even occasional takes in foreign genotypes. Two tumors gave occasional takes, but these did not reproduce as true variants after selective transfers (false positives). One tumor was nonspecific from the very beginning and continued transplantation in a foreign genotype did not result in the selection of cells particularly suited for that genotype with the exclusion of others. One tumor, finally, gave occasional takes in a foreign mouse stock, which, when transferred selectively, resulted in a line that would grow in a high percentage of that stock. These types can all be recognized in our material. Strictly strain-specific tumors are comparable to three of our tumors (DSWB, MNSB, and MSWG). Tumors giving occasional takes in a foreign genotype but nevertheless refusing to reproduce true to type on continued selective transfer in the form of specific variants can be compared to the few false positives our MSWB sarcoma gave in $A \times A.BY$ F_1 hybrids. Tumors nonspecific from the very beginning and lacking the ability to develop into specific sublines with selective compatibility are comparable to our MSWC. The behavior of this tumor is not suggestive of distinct mutations in particular genes but recalls more the behavior of many long-transplanted and "nonspecific" carcinomas and sarcomas. Such tumors were studied extensively by Hauschka and Levan (1953), who have demonstrated a correlation between the development of heteroploidy in their cell populations and the loss of strain specificity. It is possible that the behavior of MSWC is due to some such mechanism. Finally, the last category, comprising tumors accessible for the selection of specific variant lines, would be comparable to our MSWB and its variant in A.SW mice.

In conclusion, investigations on the behavior of primary tumors by tests involving the more or less indirect study of histocompatibility genes indicate that variations may exist between different tumors of identical origin with regard to their primary behavior, with regard to the changeability of this behavior, and, in cases of changeability, with regard to the possible pathways of change. It is tempting to speculate that these variations reflect differences in the cytogenetic make-up of the different cell lineages. Such differences may well be secondary consequences of the same basic neoplastic transformation, where branching into different directions would occur as part of the evolution of the established tumors. It is equally possible, however, that the neoplastic transformation may travel a number of different and not necessarily related pathways that all lead to the same end result, neoplasia, but are different with regard to the underlying cellular mechanism.

Before leaving the subject of histocompatibility and isoantigens, we should like to deal briefly with some changes in isoantigenicity that are probably cytoplasmic in nature and do not seem to involve mutations of histocompatibility genes. Again, the evidence is indirect but nevertheless suggestive. Barrett and Deringer (1950) and Barrett *et al.* (1953) have described the regular occurrence of a change in the histocompatibility requirements of a mammary carcinoma of C3H origin after passage through F_1 hybrid mice, derived by outcrossing C3H with another, resistant, strain. We studied the mechanism of this change (Klein and Klein, 1956) and came to the conclusion that it was not due to the selection of pre-existent variants but could be regarded as having been induced by some factors present in the host environment of the F_1 mouse. Available evidence suggests that this represents a case of nongenetic modification that does not seem to involve a loss of isoantigens but rather an increased tolerance to certain isoantibodies. This view is based on the observation that the difference between the original and the altered line disappears if they are compared in F_2 or back-cross hybrids preimmunized against cells derived from the strain of origin of the tumor. Another difference between this adaptive modification and the presumably genetic changes discussed above is that while the latter may involve the strongest histocompatibility locus, H-2, the former seems to be restricted to weaker isoantigenic systems. This is indicated by the observation that all tumors that have been found to undergo a modification of the Barrett-Deringer type have retained a minimum requirement of one histocompatibility factor. Probably this one factor corresponds to the H-2 system; we have consistently failed to demonstrate any Barrett-Deringer effect upon passage of mammary carcinomas of strain A origin through A $\times$ A.SW F_1 hybrids, heterozygous for the H-2 locus only. This suggests that H-2 may be impermeable to modifications of this type.

These experiments indicate that studies on the isoantigens and the histocompatibility of tumor cells, indirect as they are, nevertheless permit some tentative distinctions between nuclear and cytoplasmic changes. It seems that the systems involved may change through genic or chromosomal mutation, followed by selection, but also as a result of environmentally induced adaptation, probably cytoplasmic in nature, at least in some specific cases.

Chromosome Cytology

In this section, we should like to discuss some of the available information regarding the chromosome cytology of neoplastic tissues. Since this is a field with an enormous literature, it is not possible to cover more than a few limited aspects. We shall first deal with some recent studies on the chromosome constitution of *primary* tumors. It would be well to emphasize the word primary, as contrasted to transplanted or explanted tumor lines, which may have undergone a secondary evolution of their own. Evidence on this subject is still scarce, owing mainly to technical difficulties. There are some informative recent papers available dealing with spontaneous or induced human, animal, or plant tumors. They are supported by DNA determinations on individual nuclei. Ising and Levan (1957) have collected available critical data of recent origin regarding the modal chromosome number of different human tumors. After having pooled these with their own data and also with the DNA determinations of Leuchtenberger *et al.* (1954), they arrived at a total of 53 analyzed neoplasms out of which 16 per cent had hypodiploid modes, 58 per cent had modes between diploid and triploid, 15 per cent between triploid and tetraploid, and 11 per cent were hypertetraploid. They concluded that the optimal region for human tumors, i.e., the region within which the stemline mode of different tumors tends to gather most frequently, is between diploid and triploid. Characteristic ring chromosomes were present in two of their tumors. It is interesting to compare this information with the recent DNA measurements on human tumors by Atkin and Richards (1956). These authors found that normal leukocytes, lymphocytes, plasma cells, and fibroblasts had almost identical amounts of DNA. Nonmalignant endometrial and cervical tissues had DNA amounts that were about 10 per cent higher. There was little variation in these samples, apart from some polyploid cells. On the other hand, 17 samples of malignant tumors from 12 different sites showed considerable variation in their DNA content. The modal values of 8 tumors were from 10 to 30 per cent above the control value, while 9 tumors were grouped on both sides of the double control value. The authors conclude that the basic modal DNA value in tumors differs frequently from the normal and take this to indicate that each tumor has its stemline cell, which frequently differs from the homologous normal diploid cell in its characteristics.

Manna (1955) has examined the chromosome number of more than a thousand mitotic figures from 29 different cases of human cancerous cervix uteri. He found that 31 per cent of the tumors had modal values below diploid, 55 per cent were between diploid and triploid, and 14 per cent were between triploid and tetraploid. Cytological studies on the chromosome complement of six histologically closely similar human mammary carcinomas have been recently published by Fritz-Niggli (1956). She found that each tumor differed from all others, both with regard to chromosome numbers, variation of chromosome numbers, and nuclear morphology. She concludes that each tumor has its own characteristic cell strain. Exactly diploid cells are rare. The increase in chromosome numbers does not occur harmonically in the form of simple duplications.

Koller (1956) has studied the chromosome numbers in tumor cells from 98 effusions derived from 58 patients with various malignant conditions. In only two of the 98 fluids were the cells highly uniform with regard to chromosome numbers and morphology; the rest showed diverse variations and abnormalities in chromosome number and morphology.

There are only very few studies available on primary animal tumors with modern cytological methods. A recent note of Ford et al. (1957) is therefore of very great interest. These authors have studied a series of primary leukemias and reticulosarcomas in the mouse, some of which were spontaneous, while the majority were induced by radiation. They found that the stemline idea is broadly applicable to this material. A high proportion of the tumors studied were found to exhibit clear signs of individuality with respect to the stemline chromosome number, which was nearly always 41, 42, 43, or 44, the normal diploid number of the mouse being 40. There was considerable variation between the different tumors also with regard to the extent of variation about the stemline and with regard to the finer details of chromosome morphology. "New," short, long, or metacentric chromosomes may be present in some tumors. In the words of the authors, there were clear signs of individuality with respect to the cytogenetic constitution of each malignant cell

population as a whole. Provided that the normal ancestor cell of
these tumors is characterized by a constant chromosomal constitution, this would indicate that each neoplasm may be characterized
by an individual and possibly unique range of cell genotypes.

The chromosomal constitution of primary mouse carcinomas and
sarcomas has been studied by Sachs and Gallily (1956). They found
that while some tumors show a great variation in chromosome
number and morphology and a variety of chromosome and spindle
abnormalities, others (such as four out of seven spontaneous mammary carcinomas) showed no such variation and no such abnormalities.

An interesting study has been made on plant tumors by Partanen (1956) and Partanen et al. (1955). It was found that newly
isolated tumor strains of fern prothalli are essentially similar to
normal prothalli, i.e., they are predominantly haploid. In subsequent passages in culture they regularly change and both DNA
values and chromosome numbers indicate the appearance of heteroploid cells widely distributed between the normal range and much
higher levels. He concluded that polyploidy is neither causal to
nor even present in the initiation of this type of tumorous growth
but can be considered as a secondarily acquired characteristic.

This leads us to a brief consideration of the very extensive literature on the chromosome cytology of transplanted tumors and of
cell strains maintained in vitro. In the field of transplanted tumors
the stemline concept, developed by Makino (1952, 1956); Makino
and Kano (1953); Hauschka and Levan (1953); Hauschka (1953,
1957); Levan (1956a, 1956b); Yoshida (1956, 1957); Bayreuther
(1952); Richards (1955); and others has become well established.
It seems that each line of tumor cells contains one or a few predominant cell types that bear the major responsibility for the maintenance of the tumor. It is very exceptional to find purely diploid
stemlines in carcinomas and sarcomas, while in lymphomas and
leukemias, which seem to represent a special group in themselves,
diploid or very nearly diploid stemlines are common. Mammary
carcinomas of the mouse, which are often diploid at the spontaneous
stage, regularly change to aneuploid in the course of prolonged
transplantation, even if transplantation is strictly restricted to mem-
bers of the same inbred strain in which the tumor arose. Con-
comitantly with the change in ploidy, other changes may occur in
the biological characteristics of the tumors, such as increased growth

rate, decreased differentiation, increased homotransplantability, decreased hormone dependence, development of the ability to grow in the ascites form, etc. It is not known whether and in what way these changes are associated with the development of aneuploidy, with the exception of homotransplantability (Hauschka, 1957; Levan, 1956b); it can be said that all these changes are concomitant in a general way, while strict correlations still remain to be established. As to the change in modal chromosome number, it can be said that this does not usually occur by way of exact duplications of the diploid chromosome set; aneuploidy and not polyploidy is the rule. The aneuploid stemline is most frequently located at a level higher than the diploid number. Several tumor cell strains have been described that contain one or several characteristic marker chromosomes not found in the normal idiogram of the species but present in the majority of the tumor cells and specific for that particular line. The change in chromosome number has been found to be associated with profound structural rearrangements in the entire chromosome complement (Levan, 1956b). This is also well illustrated by the studies of Yerganian (1956) on a transplanted tumor of the Chinese hamster. This animal has only 11 easily characterized chromosome pairs and is therefore extremely favorable for such an analysis. According to Yerganian, the stemline cells in his tumor do not have a definite and replicable chromosome pattern; there seems to have occurred a random reshuffling of chromosomes as a result of perpetual nondisjunction during bipolar divisions.

Thus it would seem that lines of tumor cells kept in prolonged transplantion undergo an evolution characterized by changes in both chromosome number and structure. This evolution probably operates against the background of diverse mitotic abnormalities, continuously producing a great variety of different chromosomal types. Various kinds of selection pressure may be instrumental in modeling the population; they may be nutritional, immunogenetical, pharmacological, etc., in nature, but the most important type of selection may be the one that favors cell types more and more independent of homeostatic growth-regulating mechanisms. It may well be that the progression towards increased autonomy, usually called tumor progression (Foulds, 1954), is favored by the appearance of certain aneuploid but nevertheless viable cell types that are no longer competent to respond to the same regulatory mechanisms that may still influence their euploid ancestor to some extent.

The picture that emerges from studies on tissue cultures of mammalian cells of normal and tumorous origin is not very dissimilar to the results obtained on transplanted tumors. According to Hsu and Moorhead (1957), long-maintained cell strains, whether of normal or malignant origin, are usually heteroploid with very few or no diploids and tetraploids. Similar findings have been reported by Levan (1956c). This is very interesting, since it was found both by Hsu and Moorhead (1957) and by Tjio and Levan (1956) that primary cultures of normal tissues show a very high percentage of diploid elements. This would mean that suitable viable heteroploid cells, originally absent or in a minority, may have a selective advantage under *in vitro* conditions and would therefore tend to replace the original type. This might be related to the well-known fact that strains of normal cells have a pronounced tendency to become malignant after prolonged maintenance in tissue culture. Hsu actually points out (l.c.) that a change to malignancy seems to be associated with changes in chromosome numbers or vice versa, but he also emphasizes that more critical experiments would be needed to see whether the transition period from diploid to heteroploid actually coincides with a change in tumor-producing ability.

General Discussion

To integrate the karyological findings on primary, transplanted, and explanted tumors and their normal counterparts is an immensely difficult task. It is easy to see that no generalizations can be made and it is almost a commonplace that no particular chromosome pattern (or any other cytological characteristic, for that matter) has been found that would be specifically and regularly associated with all types of malignancy or even with one particular type. It seems questionable, however, whether there is much rationale in looking for such generalizations and such specific patterns. According to nearly general consensus of opinion, there is a multitude of causative factors that can produce neoplastic disease, alone or in combination. Such factors include radiation, chemical carcinogens, viruses, heritable factors, hormonal imbalance, etc. Very few investigators would now look for a single causal agent that would be responsible for all types of malignancy. The present discussion advances both immunogenetical and cytological evidence to indicate that each tumor is a population of cells with highly individualized biological characteristics and that permanent and true differences exist between different neoplasms even if they have been induced by the same agent in the same tissue of the same genotype. Nevertheless, it is often more or less tacitly assumed that there must be one basic cellular change that leads to tumorous growth. It is the validity of this assumption that seems to be questionable. Obviously it has to be postulated that the competence to respond to the homeostatic forces, whatever they are, which regulate cell division in the corresponding normal tissues, must be disturbed in a greater or lesser degree in all tumor cells. We know almost nothing about the way this competence is built up but have every reason to assume that the underlying cellular organization must be complex, and it is reasonable to speculate that both genetic determination and cytoplasmic differentiation are probably involved. A complex organization can be upset at many different points and it is conceivable that the responsiveness of the cell to growth-regulating factors may change in a number of different ways, both at the nuclear and the cytoplasmic level (or at any of the numerous smaller subunits included in these two main compartments) in such a way that the cell nevertheless remains viable and retains its reproductive integrity. Since the neoplastic transformation is a disease of differentiated cells of multicellular organisms that afflicts the level of higher organization itself, there seems to be no more reason to look for a single specific cytopathological change, identical in all tumors, than to look for one specific change in nerve cells that would explain all types of mental disease.

If this concept is adopted as a basis for discussion on, e.g., the chromosomal constitution of tumor cells, it cannot be argued that the existence of tumors with chromosome complements apparently indistinguishable from the normal idiogram critically proves that aneuploidy and other abnormalities must be of necessity unrelated to the neoplastic transformation in other tumors where they do occur, often to the complete elimination of euploid cells. It may very well be that the chromosome complement is intact in some tumors (although it is difficult to prove this conclusively) and that their malignant transformation was due to either point mutations or extrachromosomal changes. This does not alter the fact, however, that the *majority* of primary tumors so far studied are different in being characterized by various changes in their modal chromosome number, usually proceeding towards aneuploidy and often accompanied by concomitant structural rearrangements. To regard all these changes as secondary and unrelated to malignancy because they are not always found and do not conform to one specific pattern is no less extremistic than the earlier claim that all tumors were due to chromosomal abnormalities. On the basis of present evidence it would be more conservative to assume that changes in the chromosome complement *may* have a certain significance for the neoplastic transformation in some, not necessarily all, cases. This is indicated by the occurrence of a profound, progressive remodeling of the chromosome complement, favoring the establishment of aneuploid stemlines, often with the complete exclusion of euploid cells, which seems to be a regular accompaniment of the progressive evolution toward increased autonomy that occurs in serially transplanted carcinomas and sarcomas, even if they are maintained in their own inbred strain of origin. Similar changes have been found to go parallel with the change of normal cells to malignancy in tissue culture. The objection can be raised that the evolution of tumors in transplantation or explantation is artificial. This is admittedly so, but there is sufficient recent evidence available now to indicate that similar changes can be found frequently, although not invariably, in primary tumors which are more often aneuploid than euploid. Another objection, and a more difficult one, would be that this is merely an expression for aneuploidy in normal somatic cells, the tumor cells reflecting the chromosome constitution of their normal predecessors. Studies in which the DNA content or the chromosome number of homologous normal and neoplastic tissues has been critically compared indicate, however, that aneuploid cells are very infrequent in normal tissues although they may occur. To postulate that the frequent aneuploid tumors all originate from the infrequent aneuploid cells present in normal tissues would be the same as to say that aneuploid cells have a more pronounced tendency to become malignant than their euploid counterparts; and this would again mean that chromosomal rearrangements and deviations from the euploid complement may favor the development of malignancy. On the other hand, the fact that the chromosomal modes of primary tumors (both human and animal) so far analyzed tend to accumulate around certain particular aneuploid numbers, the region being characteristic for the species, more probably indicates that something similar to the evolution of transplanted and explanted tumors may have occurred also in some primary tumors and that the rearrangement of the chromosome complement may favor the progressive development of increasing autonomy, at least in certain cases. This is the same as to say that chromosomal changes may promote the neoplastic transformation, even if their role is not necessarily exclusive, general, or specific. Although we still know very little, it would be tempting to suggest that the neoplastic transformation may proceed along various alternative pathways in the cell, leading to individually differentiated marker characteristics in the different lineages of tumor cells, and that among these pathways chromosomal rearrangements may play a distinct although not necessarily dominating role.

References

AMOS, D. B., P. A. GORER, and Z. B. MIKULSKA. 1955. The antigenic structure and genetic behaviour of a transplanted leukosis. *Brit. J. Cancer* 9: 209–215.

ATKIN, N. B., and B. M. RICHARDS. 1956. Deoxyribonucleic acid in human tumours as measured by microspectrophotometry of Feulgen stain: a comparison of tumours arising at different sites. *Brit. J. Cancer* 10: 769–786.

BARRETT, M. K., and M. K. DERINGER. 1950. An induced adaptation in a transplantable tumor of mice. *J. Nat. Cancer Inst.* 11: 51–59.

BARRETT, M. K., M. K. DERINGER, and W. H. HANSEN. 1953. Induced adaptation in a tumor: Specificity of the change. *J. Nat. Cancer Inst.* 14: 381–394.

BAYREUTHER, K. 1952. Der Chromosomenbestand des Ehrlich-Ascites-Tumors der Maus. *Ztschr. Naturforsch.* 7b: 554–557.

CLOUDMAN, A. M. 1932. A comparative study of transplantability of eight mammary gland tumors arising in inbred mice. *Am. J. Cancer* 16: 568–630.

FORD, C. E., J. L. HAMERTON, and R. H. MOLE. 1957. The cytogenetic individuality of spontaneous and radiation-induced neoplasms in the mouse. *Proc. Am. A. Cancer Res.* 2: 202.

FOULDS, L. 1954. The experimental study of tumor progression: A review. *Cancer Research* 14: 327–339.

FRITZ-NIGGLI, H. 1956. Die Chromosomen in menschlichen Mamma-Karzinom. *Acta Unio Internationalis contra Cancrum.* 12: 623–637.

GORER, P. A. 1956. Some recent work on tumor immunity. *Advances Cancer Res.* 4: 149–186.

HAUSCHKA, T. S. 1953. Cell population studies on mouse ascites tumors. *Trans. N. Y. Acad. Sci. Ser. II.* 16: 64–73.

HAUSCHKA, T. S. 1957. Tissue genetics of neoplastic cell populations. *Canad. Cancer Conf.* 2: Academic Press, Inc., New York. Pp. 305–345.

HAUSCHKA, T. S., and A. LEVAN. 1953. Inverse relationship between chromosome ploidy and host specificity of sixteen transplantable tumors. *Exper. Cell Res.* 4: 457–467.

HSU, T. C., and P. S. MOORHEAD. 1957. Mammalian chromosomes in vitro. VII. Heteroploidy in human cell strains. *J. Nat. Cancer Inst.* 18: 463–471.

ISING, U., and A. LEVAN. 1957. The chromosomes of two highly malignant human tumors. *Acta Pathol. Microbiol. Scand.* 40: 13–24.

KLEIN, E., and G. KLEIN. 1956. Mechanism of induced change in transplantation specificity of a mouse tumor passed through hybrid hosts. *Transplant. Bull.* 3: 136–142.

KLEIN, E., G. KLEIN, and L. RÉVESZ. 1957. Permanent modification (mutation?) of a histocompatibility gene in a heterozygous tumor. *J. Nat. Cancer Inst.* 19: 95–114.

KLEIN, G., and E. KLEIN. 1956. Detection of an allelic difference at a single gene locus in a small fraction of a large tumour-cell population. *Nature, London, 178:* 1389–1391.

KOLLER, P. C. 1956. Cytological variability in human carcinomatosis. *Ann. N. Y. Acad. Sci. 63:* 793–817.

LEUCHTENBERGER, C., R. LEUCHTENBERGER, and A. M. DAVIS. 1954. A microspectrophotometric study of the desoxyribose nucleic acid (DNA) content in cells of normal and malignant human tissues. *Am. J. Pathol. 30:* 65–85.

LEVAN, A. 1956a. Chromosomes in cancer tissue. *Ann. N. Y. Acad. Sci. 63:* 774–792.

LEVAN, A. 1956b. The significance of polyploidy for the evolution of mouse tumors. Strains of the TA3 mammary adenocarcinoma with different ploidy. *Exper. Cell Res. 11:* 613–629.

LEVAN, A. 1956c. Chromosome studies on some human tumors and tissues of normal origin, grown in vivo and in vitro at the Sloan-Kettering Institute. *Cancer 9:* 648–663.

LITTLE, C. C., and L. C. STRONG. 1924. Genetic studies on the transplantation of two adenocarcinomata. *J. Exp. Zool. 41:* 93–114.

MAKINO, S. 1952. Cytological studies on cancer. III. The characteristics and individuality of chromosomes in tumor cells of the Yoshida sarcoma which contribute to the growth of the tumor. *Gann 43:* 17–34.

MAKINO, S. 1956. Further evidence favoring the concept of the stem cell in ascites tumors of rats. *Ann. N. Y. Acad. Sci. 63:* 818–830.

MAKINO, S., and K. KANO. 1953. Cytological studies of tumors. IX. Characteristic chromosome individuality in tumor strain-cells in ascites tumors of rats. *J. Nat. Cancer Inst. 13:* 1213–1235.

MANNA, G. K. 1955. Chromosome number of human cancerous cervix uteri. *Naturwissenschaften. 42:* 253.

PARTANEN, C. R. 1956. Comparative microphotometric determinations of deoxyribonucleic acid in normal and tumorous growth of fern prothalli. *Cancer Res. 16:* 300–305.

PARTANEN, C. R., I. M. SUSSEX, and T. A. STEEVES. 1955. Nuclear behavior in relation to abnormal growth in fern prothalli. *Am. J. Botany 42:* 245–256.

RICHARDS, B. M. 1955. Deoxyribosenucleic acid values in tumour cells with reference to the stem-cell theory of tumour growth. *Nature, London 175:* 259–261.

SACHS, L., and R. GALLILY. 1956. The chromosomes and transplantability of tumors. II. Chromosome duplication and the loss of strain specificity in solid tumors. *J. Nat. Cancer Inst. 16:* 803–841.

SNELL, G. D. 1948. Methods for the study of histocompatibility genes. *J. Genet. 49:* 87–108.

SNELL, G. D. 1955. Isogenic resistant (IR) lines of mice. *Transplant. Bull. 2:* 6–8.

SNELL, G. D., P. SMITH, and F. GABRIELSON. 1953. Analysis of the histocompatibility-2-locus in the mouse. *J. Nat. Cancer Inst. 14:* 457–480.

STRONG, L. C. 1929. Transplantation studies on tumors arising spontaneously in heterozygous individuals. I. Experimental evidence for the theory that the tumor cell has deviated from a definitive somatic cell by a process analogous to genetic mutation. *J. Cancer Res. 13:* 102–115.

STRONG, L. C., and C. C. LITTLE. 1920. Tests for physiological differences in transplantable tumors. *Proc. Soc. Exp. Biol. Med. 18:* 45–48.

TJIO, J. H., and A. LEVAN. 1956. The chromosome number of man. *Hereditas 42:* 1–6.

YERGANIAN, G. 1956. Discussion of A. Levan's paper. *Ann. N. Y. Acad. Sci. 63:* 789–792.

YOSHIDA, T. 1956. Contributions of the ascites hepatoma to the concept of malignancy of cancer. *Ann. N. Y. Acad. Sci. 63:* 852–881.

YOSHIDA, T. 1957. Studien über das Ascites-Hepatom. Zugleich ein Beitrag zum Begriff der cellulären Autonomie im Wachstum der malignen Geschwulst einerseits und der Individualität der einzelnen Geschwulst anderseits. *Virchows Archiv. 330:* 85–105.

5

Chromosomal Differentiation in Insects

WOLFGANG BEERMANN[1]

Structure and function of the chromosomes are usually considered from a genetic point of view. For more than half a century of descriptive as well as experimental cytology the emphasis has been on observations that demonstate the genetic constancy of the linear component of chromosome structure and on the processes of identical replication and transmission that ensure such a constancy. From a physiological point of view, however, identical reproduction is but one, and not even the chief, manifestation of chromosome activities. The chromosomes, in order to control the nature and range of the reactivity of the larger system to which they belong, have to communicate, i.e., interact materially, with the other constituents of the system, via the nuclear membrane. The nature of this interaction is still one of the major problems of biology, especially with regard to cellular differentiation.

In considering the wide range of cell phenotypes that are compatible with one and the same set of chromosomes, it is not merely nucleocytoplasmic interaction as such that has to be accounted for. The question, rather, is how a given type of cell, by interaction with its genome, interprets the genetic information in its own specific way. It is clear that some form of differential activation of the genes, in other words, functional differentiation of the chromosomes, must be involved, a view already implicit in the writings of Driesch (1894), O. Hertwig (1894), Boveri (1904), and many others.

[1] Zoologisches Institut der Universität Marburg/Lahn, Germany. Part of the work summarized in this article has been aided by a grant of Deutsche Forschungsgemeinschaft, Bad Godesberg, Germany.

Factual information concerning this point has been very scarce in the past, both from the descriptive and from the experimental side.

Mitotic chromosomes, of course, can hardly be expected to show any signs of a functional differentiation that might have existed during the interphase. In the latter stage, on the other hand, when the nucleus is metabolically active, the chromosomes as a rule are hard to identify so that visible manifestations of gene activity will not become detectable, beyond general phenomena such as the appearance of nucleoli or of chromocenters. The giant polytene nuclei of dipteran insects form a notable exception to this rule. It is the object of the following review to call attention to recent developments in the morphological study of polytene chromosomes that may provide a new and better basis for experimental work on the functional differentiation of chromosomes and genes (Beermann, 1952a, 1952b; Mechelke, 1953; Breuer and Pavan, 1955). In order to judge the significance of the phenomena to be described it is necessary first to recall some of the principles of insect development in general, as well as the facts of polyploidization and polytenization.

Some Characteristics of Insect Development

Developmental change comprises both differentiation and growth. The life cycle of holometabolic insects presents an extreme example of how the two processes may be separated in time. During the embryonic phase differentiation dominates the scene in laying down the organization of the larva; during larval life, on the other hand, there is practically no change in organization while an immense growth occurs; and, finally, during metamorphosis, growth (of the organism as a whole) ceases again and differentiation takes over to form the adult. Larval growth, in most organs and tissues, takes place without cellular multiplication, simply by increasing the size and mass of each individual cell. In such a system, the growth rate of, and the final dimensions reached by, a given cell must be controlled, to some extent, by physical factors as expressed by the surface-volume relationship. To a larger extent, however, the factors responsible for size differences seem to be of a "constitutional" nature; in other words, the size differences between cells of a different type are just another aspect of cellular differentiation. It is difficult to judge the physiological advantages involved in the evolution of such a differentiation, but one has to conclude from insect histology in general that, in systems turning over large amounts of metabolites, i.e., tissues with resorbing or secreting functions, the over-all efficiency of a few large cells exceeds that of numerous small cells of the same total mass, at least when rapid growth occurs. By abolishing mitotic multiplication of its constituents, a growing system saves both time and energy. Endomitosis probably does not seriously interrupt the metabolic activities of the cell as a whole. Furthermore, whereas cellular multiplication requires continuous readjustment of the structural relationships within an organ, especially in the case of a gland, this is not so if the cells simply grow larger and larger.

Endomitosis, Polyploidy, and Polyteny

There is no a priori reason why cellular growth should always entail repeated chromosomal reduplication in cases where mitotic division fails to occur. Oocytes, for instance, or the giant unicellular alga *Acetabularia,* grow, i.e., increase their cytoplasm considerably, without polyploidization of the nucleus. However, in contrast to differentiated somatic cells, both the oocyte and the alga represent systems to which no specific "function" can be ascribed other than to reproduce. While, with respect to this "function," greater size and rapid growth both may be of advantage, polyploidization, as is easily seen, must lead to serious complications. In a purely somatic cell line, on the other hand, there exists no such limitation as to polyploidy. Thus, if continuous cellular growth rather than cellular multiplication proves to be more effective in the development of an organ, endomitosis is substituted for mitotic distribution of the chromosomes. Furthermore, whenever the function of a growing cell involves the direct cooperation of chromosomal genes—as is likely in many types of synthetic activities—polyploidization should become a physiological necessity, too. For, in synthetically active cells, growth is simply a means to increase productivity, which, of course, would be impossible without a corresponding increase in the number of enzymatically active sites in the nucleus, i.e., multiplication of the chromosome strands.

It has been suggested that endomitotic reduplication of the chromosomes as such might be one of the causative factors of differentiation (Huskins, 1948), by shifting the ratio of gene activities in spite of constant numerical proportions. In insect development the evidence is not in favor of this hypothesis. Thus, while at least 10 cycles of chromosomal reduplication occur during growth of the larval salivary glands or the Malpighian tubules of *Chironomus,* there has been no indication of a change of function from the first to the fourth larval stage. At the onset of metamorphosis, on the other hand, when the cells of the salivary glands all seem to change their activity as judged by the properties of the secretion, this change occurs regardless of the size of the individual cell.

In most insects as well as in other organisms endomitosis leads to a complete separation of old and new chromatids subsequent to each reduplication, and as a result homologous elements will be distributed, more or less at random, all over the nucleus. In such cases, beginning at a stage of ploidy of from 32 to 64 n, the nucleus grows and as a rule begins to lobulate and ramify increasingly, as is seen, for instance, in the spinning glands and the Malpighian tubules of lepidopteran larvae. Ramification may facilitate the exchange of material between nucleus and cytoplasm, each lobe of the nucleus being physiologically autonomous in containing a more or less complete sample of all elements of the genome. In dipteran polytene nuclei, on the other hand, lobulation does not occur. Greater deviations from the spherical, or ovoid, shape would, of course, be useless if not detrimental, if any individual region of the nucleus contains genetic elements different from those found in any of the others.

As a variant of endomitosis, polytenization represents an extreme. The process of chromosome reproduction appears to be reduced to its essentials, the synthesis of new chromatids. Instead of falling apart, the strands undergo a very intimate lateral union and, with increasing polyteny, elongate continuously (see below). Pairing and elongation show a positive correlation, but it is difficult to determine which of the two is cause and which is effect. The regular occurrence of somatic pairing in dipteran nuclei that are not polytene may point to an interpretation in terms of an unusual attraction between homologues which, for mechanical reasons, forces the fibers (chromonemata) to stretch as their number increases. On the other hand, attraction itself could first result from an exceptional degree of uncoiling of the chromosomes, which would thus be typical of dipteran nuclei in general (cf. Cooper, 1938).

Growth and General Differentiation of
Polytene Chromosomes

From measurements of the nuclear volume (G. Hertwig, 1935) and of the nuclear content of DNA (Swift and Rasch, 1953) it is known that the larval salivary gland chromosomes of *Drosophila* reduplicate up to 11 times until pupation. This corresponds to a maximal degree of ploidy of $2^{1 + 11} = 4096$ n, the number of strands per chromosome pair probably being a multiple of this value (on the hypothesis that the mitotic chromatids are multistranded). Chironomid salivary gland nuclei may even reach values as high as 32,768 n ($2^{1 + 14}$), as judged by their size (Besserer, 1956) and by other criteria (Beermann, 1952a).

The chromosomal reduplication process scarcely, if ever, manifests itself visibly in polytene chromosomes. Probably as a last remnant of an endomitotic contraction cycle, slight variations in the state of contraction of the chromosomes have been observed in the earliest stages of polytenization in *Chironomus* (Beermann, 1952a). Later on, cyclic changes are no longer apparent. More or less conspicuous directional changes do, however, take place, depending on the species and tissue studied. In *Chironomus* (Beermann, 1952a) the earliest stages of polytenization are characterized by a very tight coiling of the chromosomes, which relaxes with increasing polyteny. During this stage ("spiral stage") the chromosomes appear flat and ribbonlike. This is true also for the following stage, which has been called the "meander stage" on account of the peculiar torsional curling assumed by the chromosomes. Such an appearance seems to be a mechanical consequence of the conflict between the internal torsion of the strands and the increasing width of the fibrillar sheet as a whole. On further multiplication of its strands the "meander" chromosome becomes packed with more and more coiled fibrillar sheets and gradually assumes the familiar compact, cablelike aspect. All changes described have two features in common, viz., a tendency of the constituent fibrils to coil and a tendency of the fibrils to arrange themselves in sheets. The latter point indicates a bilateral (bifacial) organization of the single strand, a consequence inherent to any double helix model of the molecular structure of the genetic strand.

The length of polytene chromosomes is primarily a function of the number of the constituent strands. This relationship becomes immediately apparent in structurally hybrid triploids in which one of the three partners of a somatic "trivalent" sometimes remains completely unpaired. It is then always shorter than the homologous "diploid" element within the same nucleus. In the fourth salivary gland chromosome of *Chironomus* the increment per each doubling thus obtained is between 15 and 20 per cent (length ratio haploid/diploid, mean for 26 cases = .82, S.E. of the mean = .022), meaning that the length attained in 14 cycles of reduplication would be more than 10 times the original one. A direct comparison of the fourth chromosome in pachytene with the fully grown homologous element in salivary glands likewise renders a ratio of about 1:10. The constitution of the animal, genetically and physiologically, the type of the cells and the chromosomes considered, and external conditions are further factors influencing the length of the chromosomes (see especially the work of Wolf, 1957) on the allocyclic behavior of the salivary X chromosome of *Phryne cincta*). In *Chironomus* (Beermann, 1952a) it has been found that polytene chromosomes of the Malpighian tubules and of the midgut are considerably longer (up to one and a half times) than those of the salivary glands, although the latter definitely reach higher values of polyteny (corresponding to four or five additional reduplications). Con- comitant to this type of intertissue variation, there are conspicuous differences in shape if polytene chromosomes from different tissues of *Chironomus* are compared, e.g., a nucleus of 25 micra in diameter may show slender, ribbonlike, coiled chromosomes in the midgut; chromosomes of the "meander" type in the Malpighian tubules; and stout, nearly cylindrical chromosomes in the salivary glands. The cause of these differences is not clear. They probably point to differences in the "initial" length of the chromosomes at the onset of polytenization. On the other hand, the length increment per reduplication cycle might in itself be subject to tissue-specific variation.

It is unlikely that the longitudinal "growth" of giant chromosomes involves changes other than submicroscopic uncoiling or unfolding, i.e., stretching. The stretching as such seems to be restricted to the so-called interbands. As a result, the concentration of genetic material in these regions is below the limit of detectability by the Feulgen reaction and the UV spectrograph. In structural terms it follows that, while the elementary fibrils must be very tightly packed (folded) within the bands, their spacing within the interbands would be wide enough to allow the free passage of the nuclear sap across the chromosome. Electron-microscope observations (see page 99) and measurements of the concentration of dry mass, by means of X-ray absorption (Engström and Ruch, 1953), further bear out this conclusion.

Chromosomal Differentiation at Specific Loci, in Relation to Cellular Physiology

Variation in the mode of polytenization and in the state of contraction has been shown to lead to a certain degree of morphological differentiation of giant chromosomes in different types of cells. This type of differentiation may be compared to the differences in nuclear structure commonly found in normal diploid cells and described as variations in the state of dispersion of the "chromatin." Although such differences may reflect variations in nuclear physiology, it is obvious that they bear no relation to the hypothesis proposed in the introduction to this paper, namely, differential "activation" of the genes in different types of cells, during and subsequent to the act of determination. The study of this basic phenomenon, however, requires a degree of precision in the morphological analysis that again only the polytene chromosomes can provide. The banding pattern of the polytene chromosome may be regarded as an enormously enlarged reflection of the longitudinal organization of the single genetic strand. Up to several thousands of individual loci can thus be exactly defined within a set of polytene chromosomes as first demonstrated by Bridges (1935).

Polytene chromosomes are found in differentiated, fully functional cells. They may therefore be expected to present different specific patterns of gene activities, depending on the type of cell investigated. Structural changes correlated to such a functional differentiation of the chromosomes ought to appear superimposed upon the genetically fixed pattern of banding as defined by the Feulgen reaction. Gene "activation," in the sense of a disproportionate increase of activity brought about by an excessive supply of substrates, might manifest itself visibly in several different ways. For instance, if one could make the gene products precipitate and accumulate at the places where they are being formed, the polytene chromosomes ought to become bloated with these products, so that the swelling of each individual locus would be a measure of its activity. Now the spontaneous occurrence of localized swellings ("bulbs") caused by the accumulation of material has been frequently observed, especially in the salivary gland chromosomes of *Chironomus* (e.g., Bauer, 1935). In the living gland, under phase contrast, the material appears in the form of droplets "sandwiched" in between two or three thicker bands (Plate V–1). The material proves to be Feulgen negative but shows positive reactions for RNA and proteins. The distribution of the loci where droplets are formed is identical throughout the nuclei of the salivary glands of a given individual (in *Chironomus*). However, there is no doubt about the physiological nature of the phenomenon as such. Its expression is highly variable from individual to individual (as stated also for *Sciara* by Poulson and Metz, 1938), and it may even be induced experimentally: Thus, when *Chironomus* larvae are transferred, first to water of 5° C for at least one hour, and then back to water of 20° C, droplets will appear regularly at a certain group of loci (see also the next chapter) after a period of from 1½ to 2½ hours. The frequent observation of droplets when material freshly collected from outdoors is being dissected is almost certainly due to this temperature effect. The precipitation of droplets as such probably derives from a temporary shift of the intranuclear balance between reaction and diffusion rates. All droplets have disappeared 8 hours after the transfer to 20° C.

There is a second and more fundamental way in which the activation of genes could manifest itself at the cytological level. By analogy to what is known of enzymatic reactions, one may postulate that the reaction between the gene and its substrates, instead of being a matter of momentous collision, requires the temporary formation of a gene-substrate complex, with structural characteristics different from those of the gene locus as such. In terms of polytene chromosomes, activation might therefore become manifest in visible alterations of the fine structure of the activated locus (band), the magnitude of the changes depending on the degree of activation, i.e., number of substrate molecules available per unit time (= probability of the formation of gene-substrate complexes).

The only reliable distinguishing character of a band, apart from its position within the banding pattern, is its content of DNA as compared to that of the others (Rudkin *et al.*, 1955). Since the concentration of the DNA, for a majority of bands, is not subject to much variation, the thickness of a band is a crude measure of its DNA content (see, however, the exceptions to follow), i.e., it may be considered as a genetic character in normal circumstances. This relationship is the basis of what is called the "constancy" of the banding pattern of polytene chromosomes (cf. Slyzinsky, 1950; Beermann, 1950, 1952a; Pavan and Breuer, 1952), notwithstanding apparent exceptions where "extrachromosomal" DNA may be involved. Different bands may also differ in the degree of "granulation," i.e., visible subdivision into smaller or larger granules. It is not clear how much of this variation is due to genetic differences because there are large differences in the general degree of granulation between chromosomes from different individuals and from different tissues. No such uncertainty exists with regard to a third source of variation, which involves the concentration of DNA within the bands, or the density of band structure.

Any preparation of polytene chromosomes shows a number of bands, which, when stained with Feulgen or with one of the usual basic stains, look "pale" and have diffuse boundaries. The same bands, on staining with Unna's methyl green–pyronin, display a high affinity to the pyronin (Brachet, 1944; Pavan and Breuer, 1955), indicating the presence of RNA. Furthermore, with fast green (or light green) at neutral pH as a counterstain after Feulgen or orcein, they appear green (Schultz, 1941; Pavan and Breuer, 1955; Swift and Rasch, 1953; and own observations, cf. Plate V–2, 3), indicating the presence of "higher" proteins (as distinguished from histones). Presence of RNA is further indicated by a metachromatic staining with toluidine blue (Beermann, 1952a).

Structurally, the "dilution" of the DNA (and probably of the histone too) is most easily understood as a sort of unfolding, or uncoiling, of the nucleoprotein fibers within the diffuse bands (and "puffs," see text following), perhaps brought about by the attachment of ribonucleoprotein. Evidence in favor of this hypothesis will be presented later (p. 95). The presence of RNA is now generally considered as being indicative of protein synthesis, so that diffuse bands, on account of their RNA content alone, may be considered as places of high synthetic activity. There are several further arguments, both direct and indirect ones, to strengthen this view. First of all, when droplet formation is induced in the salivary gland chromosomes of *Chironomus* (as previously discussed) the droplets as a rule develop within some of the diffuse bands (Plate V–1), or "puffs," i.e., bands

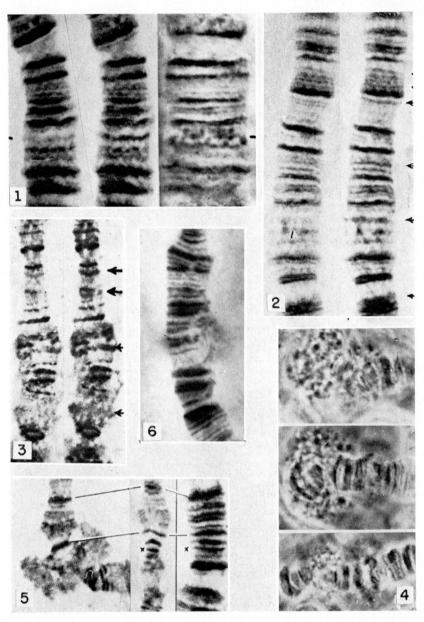

Plate V. (1) *Chironomus pallidivittatus*, salivary Chromosome 1. (Left and middle): Orcein—light green stained, photographed in green and red light, respectively. (Right): From living gland, after cold pretreatment, phase contrast, to show droplets. Droplets coincide with one of the "green" bands (arrows). (2) C. *pallidivittatus*, salivary Chromosome 1. Orcein—light green. Photographed in green and red light, respectively. A series of "green" bands is that have become swollen to such an extent that their original band nature is no longer apparent. In addition, recent studies with isotope-labeled amino acids indicate that the level of incorporation (or turnover) is highest again at certain puffed regions, e.g., Balbiani's rings in *Chironomus* (preliminary reports of Ficq, 1957, and Gross, 1957). A convincing type of argument finally comes from a comparative analysis of the state of puffing in cells of different function; this will be discussed subsequently.

Chromosomes and Cellular Differentiation: Puffs and Balbiani's Rings in Different Types of Cells

Different genes do different things. The hypothesis of differential gene activation implies, in addition, that different genes may differ quantitatively as regards the level of their activity and that the same gene may show any degree of activation depending on the type and the functional state of the cell. As shown in the preceding paragraph, puffing seems to be an expression of raised synthetic activity. Therefore one should expect to find, as a more or less reliable reflection of an underlying functional differentiation of the chromosomes, different "patterns of puffing" characteristic for each type of cell and each type of function, in a given individual. This expectation has been fulfilled (Beermann, 1952a, 1952b; Mechelke, 1953; Breuer and Pavan, 1955).

In the chironomid family of midges several species of the genus *Chironomus* (=*Tendipes*) are large enough to allow the cytological analysis of giant chromosomes in at least four different types of cells: those of the salivary glands, the Malpighian tubules, the midgut, and the rectum of the larvae. *Chironomus tentans* and *C. pallidivittatus*, a pair of sibling species, have been studied in detail

---

shown (arrows). (3) C. *tentans*, portion of a Malpighian tubule chromosome. Orcein–light green, photographed in green and red light, respectively, to show distribution of "green" bands. (4) *Chironomus tentans*, salivary Chromosome 4. Living, phase contrast, after cold pretreatment. Droplet formation on BR 1 and BR 2 (cf. Plate VI), three consecutive stages, starting from below. BR 2 is seen to be lagging behind. (5) C. *tentans*. (Left): Balbiani ring in Chromosome 1 of a nucleus from the proximal half of the Malpighian tubules. (Middle): Homologous section in distal half of Malpighian tubules. (Right): Homologous section in salivary glands. ✕ indicates band from which BR originates. Orcein–light green; red filter. (6) C. *pallidivittatus*, salivary Chromosome 3. Heterozygous exceptional Balbiani ring.

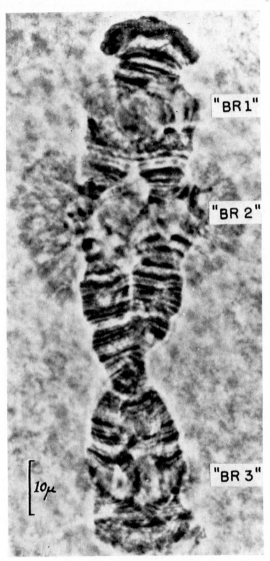

Plate VI. *Chironomus tentans.* Salivary Chromosome 4, showing the normal appearance of the three rings of Balbiani, with BR 1 forming an ordinary puff.

(Beermann, 1952a, 1955). In these species one finds up to about 10 per cent of the bands in a more or less diffuse condition, depending on the type of cell and on the individual studied. Striking and constant differences in the behavior of homologous bands, as defined by their location, have been observed in comparing chromosomes from different tissues, so that the same band may be intensely stained and sharply delimited in the cells of one organ and be "pale" and diffuse in the cells of another. In considering the chromosome set as a whole, it is not as a rule the state of the single band that is characteristic of a given type of cell. The specificity, as is to be expected, rests in the combination of loci that appear diffuse, i.e., in the "pattern of puffing."

One class of puffs, "Balbiani's rings," has been found to be particularly useful in studying the correlation between the phenomenon of puffing and cellular differentiation. Since the time of Balbiani (1881) the giant salivary gland chromosomes of chironomids have been known to develop a small number of large annular protrusions (usually not more than three), the outlines of which appear diffuse, in contradistinction to the nucleolus. Morphologically, the puff nature of these rings is not immediately apparent. In cytochemical as well as in physiological terms, however, there is a complete correspondence to ordinary diffuse bands and puffs. Large quantities of ribonucleoprotein may be demonstrated and considerable droplet formation occurs after cold pretreatment (Plate V–4). The structural peculiarities of Balbiani's ring, namely, a ringlike protrusion of the puffed region around the chromosome and splitting of the chromosome on both sides of the ring, have been interpreted as follows (Beermann, 1952a). Assume that the locus in question has been in an extremely puffed (unfolded) condition from the beginning of polytenization and throughout all the life of the larva (which has been verified back to the beginning of the second larval instar). On the hypothesis that extreme puffing leads to a complete separation of the strands along the puffed locus, one has to expect that the inherent torsion of the strands causes them to coil up separately within the puffed zone so that each forms an individual loop protruding sideways from the chromosome. This interpretation has been largely verified by electron-optical observation (Beermann and Bahr, 1954).

In the salivary glands of both *Chironomus tentans* and *C. pallidivittatus* there are three loci where Balbiani's rings ("BR's") regularly appear, all of them in the small fourth chromosome (Plate VI). In larvae collected from their natural habitats BR 2 is the largest and BR 3 the smallest of the three. BR 1 is peculiar in that it is very variable from individual to individual and from cell to cell, so that it may even completely fail to develop or be replaced by an "ordinary" puff. In larvae reared from inbred laboratory strains, on

the other hand, BR 1 is fully developed as a rule and may sometimes surpass BR 2 in size. The nature of this differential behavior
is not known. In organs other than the salivary glands the "salivary"
BR's have never been observed, i.e., the fourth chromosome exhibits
a normal banding at the loci in question. This difference is so dis-
tinctive that one might be tempted to explain it by an over-all in-
capacity of the chromosomes to form BR's in all but the salivary
gland cells. However, typical, large BR's may also develop in
Malpighian tubule chromosomes, though at locations different from
where BR's are found in salivary glands. In the species under discussion, one BR regularly occurs in the right arm of the first chromosome, in the proximal half of the Malpighian tubules, comprising
about 12 nuclei (Plate V–5). Not even a trace of puffing is found
at the same locus in the salivary glands nor, interestingly enough,
in the distal half of the Malpighian tubules themselves. The same
is true for a further BR in a near terminal position on Chromosome
4. The distal half of the Malpighian tubules, on the other hand, has
been found to be characterized, in some individuals, by a BR of its
own, situated on the third chromosome.

The situation existing in the Malpighian tubules of *C. tentans*
and *C. pallidivittatus*, namely, functional differentiation within one
and the same organ, finds a complete parallel in the salivary glands
of several chironomid genera and species. At present, seven chironomid species, belonging to the genera *Chironomus*, *Cryptochironomus*,
Trichocladius, *Acricotopus*, and *Zavrelia*, are known to possess salivary glands which are functionally subdivided so that the duct of
the gland is surrounded by a small number of large "special" cells,
often with a cytoplasmic structure visibly different from that of the
"ordinary" gland cells. The "special" cells presumably add some
component to the saliva as seen directly in *Acricotopus* (Mechelke,
1953).

With respect to general structural features such as the state of
coiling and the precision of banding, there is no difference between
the polytene chromosomes in the two types of salivary gland cells
in all of the cases studied. Spectacular and consistent differences,
however, are observed in the location of Balbiani's rings. In a
majority of cases, for instance, in *Trichocladius* (Beermann, 1952b)
and in *Acricotopus* (Mechelke, 1953), the differentiation is absolute, i.e., the two types of cells do not have a BR in common, each

being characterized by from one to three BR's of its own. In *Cryptochironomus,* on the other hand, one or two BR's may develop from the same bands in both parts of the salivary glands, whereas additional BR's appear in one or the other exclusively (Bauer, 1953).

Chromosomes and Development: Reversibility of Puffing

The observations recorded above leave little doubt that each type of differentiated cell is characterized by a specific pattern of diffuse bands, puffs, and Balbiani's rings, and it appears justified to conclude that analogous structural modifications occur in any type of interphasic chromosome, whether it is polytene or not (compare, for instance, recent reports on the fine structure of lampbrush chromosomes in amphibians by Gall, 1954, and Callan, 1957). As stated earlier, the facts easily lend themselves to an interpretation in terms of differential gene activation. The question of reversibility is of crucial importance to such an interpretation. It has been studied by comparing different developmental stages of the animal.

In speaking of chromosomes, irreversible changes are defined as mutations, i.e., changes that will be reduplicated on reduplication of the chromosomes. Changes that survive replication without being replicated are not known except on the atomic level (incorporation of isotopes). If puffing of a given locus lasted throughout the lifetime of a growing dipteran cell, this could, of course, no more be considered as proof of chromosomal mutation than the fact that somatic cells of vertebrates multiply true to type both *in situ* and in tissue culture. A crucial experiment would be to introduce "differentiated" chromosomes into the nuclei of the germ line and let them pass the test of recombination. At any rate, in relation to the phenomenon of puffing, these considerations do not apply. In chironomids (Beermann, 1952a, 1952b; Mechelke, 1953) and in sciarids (Breuer and Pavan, 1955) puffing has been demonstrated to be completely reversible, e.g., at the beginning of metamorphosis. Mechelke (1953) has found that, in *Acricotopus,* the rings of Balbiani regress completely when pupation begins, with a slightly different time schedule in the two different parts of the salivary glands. A particularly vivid picture of the dynamics of puffing is due to the work of Breuer and Pavan (1955) on *Rhynchosciara angelae.* The gregarious habits of the larvae of this species seem to induce an amazing synchrony in their development so that the situation found in one animal is representative of the group as a whole and a recording of the chromosomal changes on an absolute time scale becomes possible. A majority of the more conspicuous changes occurs when the larvae cease to feed in preparation to metamorphosis. Chromosomal regions that, throughout the earlier part of larval development, exhibited normal banding will "suddenly," within 2 or 3 days, blow up into a huge puff, starting from one inconspicuous single band and, in another 2 or 3 days, return to the original state again. Different puffs reach different maximal sizes, each at a different time. A peculiarity that finds no parallel in chironomids has not yet been sufficiently explained: some regions of the sciarid chromosomes develop a "pycnotic" appearance before they puff, others do so after puffing has regressed, and some appear pycnotic both before and after puffing. Preliminary measurements of Rudkin (1957) point to changes in the content of nucleic acids, perhaps DNA, which would then have to be classified as "nongenetic."

Genetic Aspects of Chromosomal Differentiation

The fact that, in *Drosophila*, hundreds of gene mutations are known, none of which entails changes in the cytological appearance of the loci concerned is sometimes quoted as evidence against the reliability of the puffing phenomenon in judging the functional state of a locus. On the theory of differential gene activation, however, the chance of finding a mutation in a locus which plays a dominant role in the metabolism of a salivary gland nucleus must be low if mutations are selected on criteria entirely unrelated to the physiology of the salivary glands. Some of the larval lethals may prove to be more promising in this respect (e.g., lgl and lme, see Hadorn, 1955). Furthermore, on the hypothesis that puffing is an expression of quantitative rather than qualitative changes in activity, a great number of mutations, namely those that do not change the substrate specificity of the genes, would have to remain undetected at any rate. This may be one of the reasons why heterozygosity in the development of diffuse bands, puffs, and rings of Balbiani has been observed so rarely, even in species hybrids (*C. tentans* × *C. pallidivittatus*, Beermann, 1955). The first observation on "structural heterozygosity" of this kind is due to Hsu and Liu (1948) who found, in the case of a so-called "bulb," that the two types of homozygotes were in equilibrium with the heterozygotes in a Chinese population of *Chironomus*. In *C. tentans* the rings of Balbiani of the salivary glands (BR 1 and BR 2, cf. Plate VI) may appear heterozygous in some of the larvae collected from nature. An exceptional BR, which develops on the third salivary chromosome both in *C. tentans* and *C. pallidivittatus*, has also been seen to develop heterozygously in some animals (Plate V–6). In these cases, there is no detectable structural heterozygosity at homologous locations in the Malpighian tubule chromosomes, as expected. It is quite clear that future genetic work on the phenomenon of puffing will have to concentrate on finding and locating mutations that change the metabolism of such types of cells as possess polytene chromosomes of sufficient cytological quality.

Fine Structure of Puffed Regions

The foregoing discussions have been based on the assumption that puffing consists in an alteration of the structural state of the constituent fibers of the chromosome, within the confines of a band. More specifically, this alteration is thought to involve uncoiling or unfolding of the chromomeric sections of the chromonemata, which, ordinarily, would be coiled up tightly, as first visualized by Ris (Ris and Crouse, 1945). The hypothesis does not exclude the possibility that the chromomeric sections of the genetic strands differ in their structural make-up from the interchromomeric ones. Uncoiling of the chromomere sections is thought to be caused by the formation of temporary chemical complexes between substrate molecules and the active section of the genetic strands within a chromomere. Direct evidence for the uncoiling hypothesis of puffing comes from light- and electron-microscope studies of Balbiani's rings in the salivary glands of *Chironomus*.

The puffed peripheral zone of Balbiani's ring as a rule looks completely homogeneous in the light microscope, both in the living gland and in acetocarmine squash preparations. Very rarely it gives the impression of being composed of diffuse threads of a thickness close to the limit of resolution. This impression is confirmed by the electron microscope, using ultrathin sections of salivary glands fixed in Palade's OsO₄ (Beermann and Bahr, 1954). Each individual thread, i.e., that portion of each chromonema which is puffed, has the appearance of bent cylindrical "brush" of 0.2 μ diameter. The axial filament of these brushes has not been discernible so far.[2] It seems to measure less than 100 Å in diameter. The "bristles" are thicker, from 100 to 150 Å in diameter. Spherical particles of about 300 Å diameter are found to be attached in great numbers to the "bristles" all over the puffed sections of the chromonemata. These do not seem to be a necessary attribute of all BR's, since they have not been detected in ultrathin sections of Malpighian tubule nuclei (Beermann, unpublished observations) where BR's are also present (see p. 96). This argument is based on the observation that some of the granules are always seen to be distributed throughout the nuclear sap in salivary gland nuclei, and also in nuclei of the midgut (unpublished).

The length of the puffed section of the individual chromonema within Balbiani's ring is of particular interest with regard to current views on gene structure. The puffed sections form loops, and the length of these loops has been estimated, for one of the large salivary BR's of *Chironomus*, to be of the order of 5 micra. As compared to the thickness of the single band from which the loops can be shown to develop (Beermann, 1952a), this is at least a fivefold elongation, a fact which leaves little doubt that, in normal conditions, the chromomeric regions of the chromonemata must be coiled up tightly, as postulated above. Similar conclusions have been reached with regard to the loops of amphibian lampbrush chromosomes (cf. Callan, 1957). In spite of their length, the puffed sections of the strands must be regarded, on account of their uniform behavior, as functional units. The situation thus provides a new morphological basis for the apparent contradiction between the functional and the cytogenetical definitions of the "gene." As is well known, mutations which, functionally, behave as alleles, i.e., show allelic interaction in the heterozygote, often prove to be nonallelic on the criterion of recombination ("pseudoallelism," Lewis, 1951; Pontecorvo, 1956). To the morphologist it has often been difficult to visualize these relations, in view of the postulated smallness of the "gene."

---

[2] An alternative interpretation of the brushlike appearance may be based on the assumption that what is seen as "bristles" are sections of a loosely coiled and folded continuous strand.

Conclusions: The Hypothesis of Differential Gene Activation

The phenomenon of differential puffing as observed in the polytene giant chromosomes of dipteran insects has been interpreted and described in terms of the hypothesis of differential gene activation. In order to become amenable to experimental test the main points of this hypothesis require some specification.

"Activation" as a process has been defined, in its most simple form, as an increase in activity caused by an increase in the availability of one or more substrates. Other modes of "induction" (in enzymological terminology) are conceivable, e.g., by the action of a specific inducer, or by removal of a specific inhibitor, both of which would not participate in the reactions catalyzed by the gene. Such mechanisms might be invoked in order to explain the stability of the functional state of a gene independent of short-term variations in cellular function, e.g., secretion cycles. It is doubtful, however, whether the interrelations between cellular metabolism and nuclear activities are of such a direct nature as is implied in the argument. At any rate, as long as so little is known about nuclear metabolism, especially about the nature of gene substrates and gene products, the experimental distinction between different modes of activation will be difficult.

What is the meaning of "differential" in the term *differential activation?* Not more than 10 per cent of the bands of polytene chromosomes have been found to be in a diffuse, or puffed, state. Frequently the proportion of puffed loci is much lower. It would be careless to conclude from these observations that, in polytene nuclei, the large majority of genes were completely inactive or "dormant" (Beermann, 1952a, 1952b). In discussing the cytological evidence, the emphasis has to be placed on the specific differences observed between the pattern of puffing of homologous chromosomes from different types of cells. These, in the opinion of the author, show conclusively that there are large differences in the spectrum of gene activities. Whether or not all genes take part in such a differentiation, and whether or not there are nonfunctional loci, and how many, remains to be determined by experiment, e.g., by the induction of genetic mosaics involving chromosomal deficiencies (cf. Demerec, 1934). The nonoccurrence of puffing as such cannot be considered as evidence for nonactivation. Visible puffs may represent only a small fraction, though probably the dominant one, of those loci that control nuclear and cellular metabolism in a given case.

It remains to be stated that the entire discussion on chromosomal differentiation, physiologically as well as morphologically, relates to the so-called "euchromatic" portions of the genome. Puffing has never been observed in "heterochromatic" sections of polytene chromosomes. The question of the nature of the "heterochromatin" is still one of the mysteries of cytogenetics and cellular physiology.

References

BALBIANI, E. G. 1881. Sur la structure du noyau des cellules salivaires chez les larves de *Chironomus*. *Zool. Anz. 4:* 367–641.

BAUER, H. 1935. Der Aufbau der Chromosomen aus den Speicheldrüsen von *Chironomus Thummi* Kieff. *Z. Zellforsch. 23:* 280–313.

BAUER, H. 1953. Die Chromosomen in Soma der Metazoen. *Zool. Anz. Suppl. 17:* 252–268.

BEERMANN, W. 1950. Chromomerenkonstanz bei *Chironomus*. *Naturwiss. 37:* 543–544.

BEERMANN, W. 1952a. Chromomerenkonstanz und spezifische Modifikationen der Chromosomenstruktur in der Entwicklung und Organdifferenzierung von *Chironomus tentans*. *Chromosoma 5:* 139–198.

BEERMANN, W. 1952b. Chromosomenstruktur und Zelldifferenzierung in der Speicheldrüse von *Trichocladius vitripennis*. *Z. Naturforsch. 7b:* 237–242.

BEERMANN, W. 1955. Cytologische Analyse eines Camptochironomus-Artbastards. I. Kreuzungsergebnisse und die Evolution des Karyotypus. *Chromosoma 7:* 198–259.

BEERMANN, W., and G. BAHR. 1954. The submicroscopic structure of the Balbianiring. *Exper. Cell. Res. 6:* 195–201.

BESSERER, S. 1956. Das Wachstum der Speicheldrüsen- und Epidermiskerne in der Larvenentwicklung von *Chironomus*. *Biol. Zentralbl. 75:* 205–226.

BOVERI, T. 1904. *Ergebnisse über die Konstitution der chromatischen Substanz des Zellkerns*. Gustav Fischer Verlagsbuchhandlung, Jena.

BRACHET, J. 1944. *Embryologie chimique*. Masson et Cie, Paris.

BREUER, M. E., and C. PAVAN. 1954. Salivary chromosomes and differentiation. Proc. 9th Intern. Congr. Genet. 1953. *Caryologia 6* (Suppl.): 758.

BREUER, M. E., and C. PAVAN. 1955. Behavior of polytene chromosomes of *Rhynchosciara angelae* at different stages of larval development. *Chromosoma 7:* 371–386.

BRIDGES, C. B. 1935. Salivary chromosome maps. *J. Hered. 26:* 60–64.

CALLAN, H. G. 1957. Paper read at UNESCO conference on "Patterns of Organisation," Edinburgh. (In press).

COOPER, K. W. 1938. Concerning the origin of polytene chromosomes. *Proc. Nat. Acad. Sci. U.S.A. 24:* 452–458.

DEMEREC, M. 1934. The gene and its role in ontogeny. *Cold Spring Harbor Symp. Quant. Biol. 2:* 110–115.

DRIESCH, H. 1894. *Analytische Theorie der organischen Entwicklung*. W. Engelmann, Leipzig.

ENGSTRÖM, A., and F. RUCH. 1951. Distribution of mass in salivary gland chromosomes. *Proc. Nat. Acad. Sci. U.S.A. 37:* 459–461.

FICQ, A. 1957. Report of J. Brachet at the UNESCO conference on "Patterns of Organisation," Edinburgh. (In press).

GALL, J. 1954. Lampbrush chromosomes from oocyte nuclei of the newt. *J. Morph.* 94: 283–352.

GROSS, J. D. 1957. Report of C. H. Waddington at the UNESCO conference on "Patterns of Organisation," Edinburgh. (In press).

HADORN, E. 1955. *Letalfaktoren.* Georg Thieme Verlag, Stuttgart.

HERTWIG, G. 1935. Die Vielwertigkeit der Speicheldrüsenkerne und Chromosomen bei *Drosophila melanogaster. Z. Vererb. Lehre* 70: 496–520.

HERTWIG, O. 1894. *Allgemeine Biologie,* 2d ed. Gustav Fischer Verlagsbuchhandlung, Jena, 1906.

HSU, T. C., and T. T. LIU. 1948. Microgeographic analysis of chromosomal variation in a Chinese species of *Chironomus. Evolution* 2: 49–57.

HUSKINS, L. 1948. Chromosome multiplication and reduction in somatic tissues, their possible relation to differentiation, reversion, and sex. *Nature* 161: 80–84.

LEWIS, E. B. 1951. Pseudoallelism and gene evolution. *Cold Spring Harbor Symp. Quant. Biol.* 16: 159–174.

MECHELKE, F. 1953. Reversible Strukturmodifikationen der Speicheldrüsenchromosomen von *Acricotopus lucidus. Chromosoma* 5: 511–543.

PAVAN, C., and M. E. BREUER. 1952. Polytene chromosomes in different tissues of *Rhynchosciara. J. Hered.* 43: 152–157.

PAVAN, C., and M. E. BREUER. 1955. Differences in nucleic acid content of the loci in polytene chromosomes of *Rhynchosciara angelae,* according to tissues and larval stages. Symp. "Cell Secretion," Belo Horizonte, Brazil.

PONTECORVO, G. 1956. Allelism. *Cold Spring Harbor Symp. Quant. Biol.* 21: 171–174.

POULSON, D. F., and C. W. METZ. 1938. Studies on the structure of nucleolus forming regions and related structures in the giant salivary gland chromosomes of Diptera. *J. Morph.* 63: 362–395.

RIS, H., and H. CROUSE. 1945. Structure of the salivary gland chromosomes of Diptera. *Proc. Nat. Acad. Sci. U. S. A.* 31: 321–327.

RUDKIN, G. T., J. F. ARONSON, D. A. HUNGERFORD, and J. SCHULTZ. 1955. A comparison of the ultraviolet absorption of haploid and diploid salivary gland chromosomes. *Exper. Cell Res.* 9: 193–211.

RUDKIN, G. T. 1957. Personal communication.

SCHULTZ, J. 1941. The evidence of the nucleoprotein nature of the genes. *Cold Spring Harbor Symp. Quant. Biol.* 9: 55–65.

SLYZINSKI, B. M. 1950. *Chironomus* versus *Drosophila. J. Genet.* 50: 77–78.

SWIFT, H., and RASCH, E. M. 1953. Quoted by M. Alfert in Composition and structure of giant chromosomes. *Intern. Rev. Cytol.* 3: 170–171 (1954).

WOLF, B. E. 1957. Temperaturabhängige Allocyklie des polytänen X-Chromosoms in den Kernen der Somazellen von *Phryne cincta. Chromosoma* 8: 396–435.

6

Changes in Nucleoli Related to Alterations in Cellular Metabolism

HANS F. STICH [1]

The morphology and chemistry of nucleoli have received a great deal of attention in recent years, as shown by the reviews of Vincent (1955) and Stich (1956a). The correlation between nucleolar activity and synthetic processes of the cell, as well as the extremely high turnover rates of RNA as compared to the stability of chromosomal DNA, demand serious consideration in any discussion of normal and pathological cellular function. The present paper will be restricted to the writer's investigations on the nucleoli of the salivary glands of *Chironomus* and of the unicellular and uninucleated *Acetabularia*. Both these subjects were selected because they offered unique advantages for analytic and experimental work. It is hoped that the results obtained on *Chironomus* and *Acetabularia* will be transferable to the conditions of other cells and might add to a better understanding of the nuclear-nucleolar and cytoplasmic relationships.

Heterogeneity of Nucleolar Substances

Nucleoli originate on specific regions of chromosomes, which may or may not be characterized by a lack of DNA or by heterochromatic regions (Vincent, 1955; Stich, 1956a). This concept, originally suggested by Heitz and McClintock, has proved valid for

[1] Associate Professor of Cancer Research, Department of Cancer Research, University of Saskatchewan and Saskatchewan Research Unit of the National Cancer Institute of Canada. The author is indebted to J. Hammerling, J. Brachet, and H. Chantrenne for supplying *Acetabularia* stocks; to D. M. Angevine for the use of the X-ray absorption apparatus built by funds made available by the Atomic Energy Commission (AT 11–1–64, Proj. 8); and to M. McDonald for a purified ribonuclease.

most animal and plant cells. The intimate morphological connection of nucleoli to chromosomes has been given as evidence that the nucleolar substances are accumulated chromosomal products, which, perhaps, transfer information to the cytoplasm of the cell. If the nucleoli are involved at all in this transmitting function, a high diversity and complexity of nucleolar substances should be expected. Investigations of this were started by applying cytochemical methods, although the chances of detecting any differences in chemical composition with these techniques were remote.

The nuclei of *Chironomus* sp. salivary gland were chosen for study because the nucleoli are relatively large and their loci of attachment to the chromosomes readily distinguished, thus making the identification of specific nucleoli possible. The data to be described will be restricted to an analysis of the nucleolar substances accumulating on particular parts of the chromosomes described as "puffs," "bulbs," or "Balbiani rings." No data concerning the chemical nature of these substances are available at present, although the morphological alterations of these chromosomal regions have been described in detail (Beermann, 1956). These compounds will be referred to as nucleolar substances, since any differential classification would be premature at present and might lead to confusion.

The nucleolar substances attached to "puffed" chromosome parts of the smallest chromosome of *Chironomus* will be used as an example. The studies were performed on chromosomes that were first, as a rule, fixed in Carnoy solution, except for group 4 below, where 10 per cent formalin was preferred, and then isolated, using fine needles. This isolation procedure was necessary to avoid contamination by particles of cytoplasm or nuclear sap, which would interfere with the cytochemical reactions. The cytochemical tests applied were as follows: (1) for DNA, the Feulgen reaction; (2) for RNA, staining with gallocyanine, toluidine blue, or azure B, all combined with RNase treatment (Brachet, 1953); (3) for protein, the Millon reaction; (4) for protein amino groups, the fast green stain (Alfert and Geschwind, 1953) and as a check blocking the amino groups by applying the method of Monné (1950).

Reference to Plate VII, 7–10 will show better than a lengthy description the differences between the two nucleolar substances. The data indicate that the nucleolar substances at the top of the chromosome possess RNA and proteins containing basic groups, whereas the nucleolar substance at the base of the chromosome has

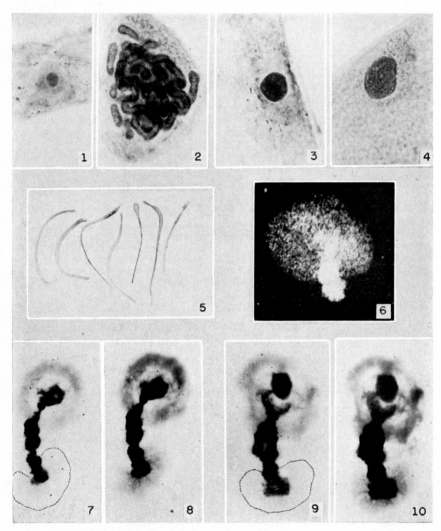

Plate VII. Nuclei of *Acetabularia mediterranea:* (1) at beginning of the experiment; (2) after 10 days' illumination; (3) after 10 days' illumination + 2,4-dinitrophenol (1:12,000); and (4) after 10 days' illumination + moniodic acid (1:10,000) (Stich, 1956a). (5) Enucleated cytoplasmic fragments of *Acetabularia crenulata* treated with RNase (0.1 mg/ml) for 4 days and kept for the following 94 days in a normal culture medium. The differentiation of any "whorls and caps" is suppressed. (6) X-ray absorption picture of an isolated chromosome with one attached nucleolus from a salivary gland cell (*Chironomus*). (7)—(10). The "small" chromosome with two attached nucleolar substances of salivary gland cells (*Chironomus*): (7) toluidine blue; (8) iron-hematoxylin stain of the same chromosome; (9) fast green; and (10) Millon reaction on the same chromosome. The outlines of the unstained nucleoli of (7) and (9) are drawn in with dotted lines.

neither RNA nor basic groups in a cytochemically detectable amount. Whatever the precise nature of the proteins may be, there is no doubt that at different loci on the chromosome substances of different chemical composition accumulate.

In a second attempt at this problem, differences in the metabolism could be demonstrated by means of the autoradiographic technique (Pelc and Howard, 1952) and through determination of the amount of organic matter in nucleoli by the X-ray absorption technique (Engström, 1946; Engström and Lindström, 1950; Clemmons et al., 1954; Clemmons, 1955). The steps necessary to complete the measurements were performed in the following order:

1. The fourth instar larvae of Chironomus were placed for 6 hours in a culture medium containing P^{32}.
2. The salivary glands were excised and the chromosomes isolated from the fixed cells (Carnoy fixative).
3. The chromosomes were transferred to prepared, special discs for the X-ray apparatus.
4. X-ray pictures were taken of the chromosomes and the content of organic matter was calculated by the microphotometric method described by Clemmons (1955).
5. The chromosomes with their nucleoli were transferred to microscope slides and covered with stripping film (Kodak) for the autoradiographic procedure. The amount of incorporated P^{32} was calculated by counting the number of grains in the emulsion as described by Taylor et al. (1955).

Plate VII–6 shows an example of an X-ray picture of a chromosome with one attached nucleolus, which was used to determine the organic matter in the nucleolar substance. The amount of organic matter thus measured was later used as a reference point for the isotope studies. The differences in incorporation and retention of P^{32} in both of the nucleolar substances can be seen from Fig. 16. The behavior of the nucleolar substances at the top of the chromosomes resembled that found in nucleoli of Chironomus Thummi (Stich, 1956a; Stich et al., in press), of Drosophila (Taylor et al., 1955), and Acetabularia mediterranea (Stich and Hämmerling, 1953; Hämmerling and Stich, 1956a, 1956b), whereas the incorporation of P^{32} into the nucleolar substance at the base of the chromosome followed another pattern.

Finally, both nucleolar substances react differently to an environmental rise in temperature from 4° C to 20° C. The nucleolus at the top decreases in size, whereas the nucleolar substance on the base of the chromosome shows a marked increase in size over a period of 2 days.

In summary, there seems to be evidence enough to support the assumption that at different parts of the chromosome substances accumulate with different chemical composition, different metabolic

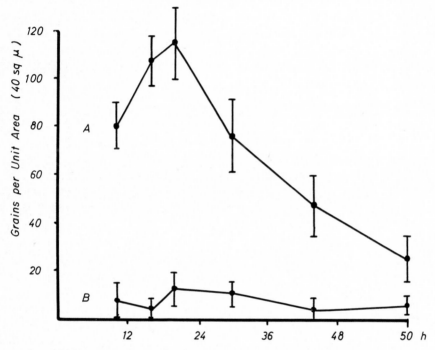

Fig. 16. Incorporation and retention of P^{32} in nucleolar substances attached at the top (curve A) and the base (curve B) of the "small" chromosome shown on Plate VII, 7—10. The fourth instar larvae of *Chironomus* were placed in a culture solution containing P^{32} for 6 hours, then transferred to a normal culture medium and fixed at the different time intervals shown on abscissae.

activity, and different response to external stimuli. At present it is difficult to decide whether these substances are collected or produced at specific loci of the chromosome; nor can any clear statement be made in regard to the role of the "puffing" of the chromosome regions and the origin of nucleolar substances, although it seems likely that the increase in these nucleolar substances causes the "puffed" modification by a spreading of the longitudinal chromosomal structures.

Regulation of Nucleolar Metabolism

The size, shape, and amount of organic matter of nucleoli vary considerably in cells of different tissues and also at different developmental periods in the life of a cell. However, from the large variety of nucleolar appearances, one commonly accepted generalization has emerged: cells characterized by a high rate of protein synthesis possess large, hypertrophic nucleoli with a high RNA content, while cells with a low rate of protein synthesis have small, undeveloped nucleoli. These findings led finally to the hypothesis proposed by Caspersson and co-workers (1950), that the protein-synthesizing system in cytoplasm depends on RNA-system function of the nucleolus. The question as to what regulates the activity of the nucleolar substances and, consequently, the nucleus remains unanswered. Thus it appeared that the uninuclear giant-cell *Acetabularia* would be extremely useful in yielding some insight into this problem and for extending the scheme of Caspersson.

The first experiment was designed to determine whether or not cytoplasm exerts any influence on nuclear and nucleolar volume. The cells selected for study were 4 mm and 8 mm in length and were taken from a culture that had been kept in darkness for a long period of time. The nuclei and nucleoli of both cell types were small and of almost equal size. The experiment consisted of first illuminating both cell types under the same environmental conditions and then of measuring the increases in nuclear and nucleolar volume. The difference in response of the two cell types, as seen from Table 6, can be attributed solely to different amounts of cytoplasm, since all other conditions were exactly alike.

TABLE 6

Effect of Amount of Cytoplasm on Nuclear Response to Illumination
in *Acetabularia* (Stich, 1956b)

	Cell Length	Initial	Four Days Light
Volume of nucleus	4 mm	27.9 ± 5.1	81.4 ± 14.4
	8 mm	30.0 ± 5.7	158.3 ± 24.1
Volume of nucleolus	4 mm	3.0 ± 1.3	6.1 ± 1.7
	8 mm	3.5 ± 1.2	12.7 ± 2.3
Number of nuclei	4 mm	1	1–4
	8 mm	1	12–19

In a second experiment an attempt was made to determine the effect of photosynthetic activity of cytoplasm on alterations of nuclei and nucleoli. In this study the volumes of the cells selected were equal, while the amount of light applied per day for various groups of cells varied from 2 to 24 hours. The increase in nuclear and nucleolar volumes was found to be directly proportional to the amount of light per day to which the cells had been exposed. Since it is only the cytoplasm with its chlorophyll-containing plastids that is able to utilize light as an energy source, the differences in nuclear responses can be attributed to different amounts of energy-rich substances formed in the cytoplasm. This assumption was supported by the finding that polyphosphates, which are long chain molecules containing high-energy phosphate bonds (Thilo *et al.*, 1956), increased proportionally with increases of the daily amount of applied light and also the corresponding increases of nuclear and nucleolar volume (Stich, 1956b).

TABLE 7

Effect of Polyphosphate Poisons and Acriflavine on Nucleus and Growth in Young *Acetabularia* 8 mm Long (Stich, 1956b)

	Initial	Controls, 10 Days	2,4-dini-trophenol, 10 Days	Moniodic Acid, 10 Days	Acriflavine, 10 Days
Volume of nucleus ...	13.2 ± 3.3	243.5 ± 39.1	57.5 ± 12.8	59.6 ± 10.1	197.7 ± 30.4
Volume of nucleolus .	0.8 ± 0.2	129.9 ± 38.8	6.7 ± 1.4	6.3 ± 1.3	79.7 ± 12.7
Number of nucleoli ..	1	18–26	1	1	10–19
Growth, mm	–	5.4 ± 2.7	0	0	0

In a final study 2,4-dinitrophenol (1:12,000) and moniodic acid (1:10,000), which are known to interfere with energy-rich phosphates, were administered to the cells of *Acetabularia* and their influence on cellular growth, protein synthesis, and nuclear changes was studied. Two different cell types were submitted to the influence of these cell poisons: (1) young cells, 8 to 10 mm long, which were characterized by a rapid growth of nuclei and nucleoli (Table 7), and (2) older cells, 2.8 to 3.0 cm long, that had fully developed nuclei and nucleoli (Table 8). In experiments using P^{32} it was shown

TABLE 8

Effect of 2,4-dinitrophenol, Darkness, and Acriflavine on Nucleus, Polyphosphate Metabolism, Growth, and Protein Synthesis in Mature *Acetabularia* 2.5 cm Long

	Control	2,4-dini-trophenol, 18 Days	Darkness, 30 Days	Acriflavine, 18 Days
Volume of nucleus	205.9 ± 38.1	72.5 ± 23.8	65.1 ± 18.8	183.0 ± 40.1
Volume of nucleolus	133.0 ± 46.4	35.2 ± 15.7	36.4 ± 19.1	129.4 ± 37.5
Number of nucleoli	9–20	1	1	8–17
P^{32} incorporation into polyphosphates	+	–	–	+
Growth	+	–	–	–
Protein synthesis ° ..	+	–	–	–

° Protein synthesis measured as increase of nitrogen in 10 per cent trichloracetic acid precipitate (Johnson, 1941).

that both compounds completely blocked the utilization of polyphosphates and led to an extreme reduction of the amount of polyphosphates in the cytoplasm (Stich, 1955, 1956b). Further, it had been shown earlier that cells of the same size and age treated with acriflavine, which combines with DNA (Wagner-Jauregg, 1943) and RNA (Stich, 1951b) *in vivo* and *in vitro*, showed no inhibition of polyphosphate metabolism before the later stage of respiratory interference (Stich, 1953; Chantrenne–Van Halteren and Brachet, 1952). If the function of nuclei depends on the synthesis of polyphosphates by the cytoplasm, it would be expected that 2,4-dinitrophenol and moniodic acid would produce nuclear changes, and by comparison acriflavine would not. If the results that were obtained and summarized in Tables 7 and 8 and Plate VII, 1–4 are considered, it can be seen that an inhibition of the energy-rich phosphates in the cytoplasm induces a marked alteration in the nuclear and nucleolar morphology. The rate of growth of young nuclei is suppressed (Table 7), while the volumes of fully grown nuclei and nucleoli are reduced (Table 8). Comparable results have been achieved by measuring the incorporation of P^{32} into the nucleoli (Hämmerling and Stich, 1956b). As shown in Fig. 17, 2,4-dinitrophenol and the prevention of photosynthesis by darkness both inhibit the incorporation of P^{32} into the nucleoli, whereas acriflavine does not.

The alterations of nucleolar morphology and metabolism, as just described, are completely reversible if the treated cells are transferred to normal culture conditions. The increase and decrease of nucleolar volume was due to a retention or loss, respectively, of organic matter and is not due to changes in water content. Preliminary studies using the X-ray absorption technique (Engström,

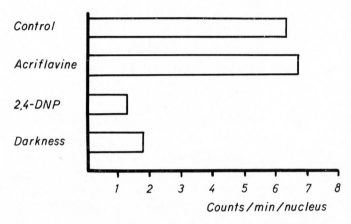

Fig. 17. Incorporation of P^{32} into the nucleus of Acetabularia mediterranea under various experimental conditions (Hämmerling and Stich, 1956b).

1946; Clemmons, 1955) have shown that the organic matter content of nucleolar substance and of nuclear sap can be reduced by 80 to 90 per cent of the original value if photosynthesis is prevented by culturing the cells in darkness (Stich, 1955). In the course of the experiment it was shown photometrically with gallocyanine stain that the nucleoli lost 70 to 82 per cent of their RNA content. Although the accuracy of the technique applied may be questioned, it can be concluded that alterations in amount of nucleolar proteins and RNA, which represent only a small fraction of the nucleolar organic matter (approximately 3 to 6 per cent), correspond to the changes in nucleolar volume.

The data just described suggest that the size of nucleoli depends on the amount of energy-rich phosphates produced in the cytoplasm. Since these compounds under normal conditions also evoke a high rate of protein synthesis in the cytoplasm, a positive correlation of nucleolar volume and cytoplasmic activity is suggested. It seems likely that cytoplasmic activity and nuclear function are kept in a state of balance by one common factor, namely, the amount of energy-rich phosphates in the cytoplasm. However, this linkage is perhaps broken in pathological conditions. For example, the cells treated with acriflavine (Tables 7 and 8, Plate VII, 1–4) have large hypertrophic nuclei with fully developed nucleolar substance, but they are unable to grow or synthesize any significant amount of protein in the cytoplasm. A second example, which is related to the present argument, is that protein synthesis in cytoplasm is "independent" of the nucleus, indicating the absence of any direct interrelationship of the two (see next section). The interpretation of a hypertrophic nucleolus as a sign of high protein synthesis in the cell as proposed by Caspersson may be valid for many cases but can be misleading when applied to cells under abnormal conditions. It might be more precise to correlate a hypertrophic nucleolus with a high amount of energy-rich compounds in the cytoplasm.

Nuclear and Nucleolar Influences in Cytoplasmic Processes

Studies on grafts between various species of *Acetabularia* and of *Acicularia* gave highly suggestive evidence for a dependence of cytoplasmic processes on the nucleus. The type of differentiation is entirely controlled by the nucleus or nuclei present in the cell (Hämmerling, 1934, 1946, 1953; Werz, 1955). The existence of "morphogenetic substances" carrying information from the nucleus to the cytoplasm has been postulated. However, any analysis of the chemical nature of these nuclear compounds released into the cytoplasm has offered the greatest of technical difficulties and no conclusive results have been obtained as yet. Some insight into the complexity of the nuclear functions might be gained by comparing the biochemistry and the behavior of nucleated with enucleated cell fragments of the same developmental stage.

In general, cytoplasm that is deprived of a nucleus survives only a short time. However, contrary to the behavior of most animal and plant cells, the cytoplasm of *Acetabularia* lives for a surprisingly long period of time without a nucleus. Enucleated cell fragments, 2 to 3 cm in length, are able to survive for as long as 10 months. They continue to grow and to form specific cell structures (Hämmerling, 1934, 1943). Growth was not due to mere osmotic swelling of the cytoplasm present at the time of enucleation but was found to be due to a net synthesis of new protein (Vanderhaeghe, 1954; Brachet and Chantrenne, 1951, 1952, 1956; Brachet *et al.*, 1955;

Stich and Kitiyakara, 1957), RNA (Brachet and Chantrenne, 1956; Vanderhaeghe and Szafarz, 1955), and energy-rich phosphates (Stich, 1953, 1956b). The amount of newly formed protein in enucleated cell pieces of a fully grown *Acetabularia*, which can exceed three to four times the original amount, is controlled by cytoplasmic factors and seems to be completely independent of the nucleus (Stich and Kitiyakara, 1957), indicating once more the autonomy of the protein-synthesizing mechanisms in the cytoplasm.

The data available at present justify the conclusion that the nucleus exerts no direct influence on these basic cytoplasmic processes. A remote control on growth, differentiation, protein synthesis, and RNA synthesis cannot, however, be denied, since all these processes cease after a period—generally in 2 to 4 weeks—following enucleation. These results immediately raise the question of the mechanism of the remote nuclear function. By analyzing this problem, it is hoped to find a better understanding of nuclear and nucleolar function.

Some interesting results in regard to this question have been achieved by applying RNase to living nucleated and enucleated cell fragments of *Acetabularia* (Stich and Plaut, 1958). The cells were cut and enucleated while in a solution containing RNase, so as to facilitate the penetration of the RNase molecules into the interior of the cell. The results summarized in Figs. 18–21 showed that (1) the RNase (0.1 mg/ml) blocks protein synthesis and growth in nucleated as well as in enucleated cell fragments and (2) the nucleated cells are able to resume growth and protein synthesis, while the enucleated pieces cannot recover after RNase treatment. The latter were still alive 3 to 4 months after enucleation but they lacked any considerable increase in protein content, growth, or differentiation (Plate VII–5).

One approach to the study of the mechanism of nuclear function consisted in adding a normal nucleus to an RNase-treated cell fragment that had lost its ability for protein synthesis and growth. It was expected that if the nucleus were involved in protein synthesis, the cytoplasm would recover all the activities stopped by the RNase treatment. The experiment was carried out as follows:

1. RNase was applied to 1.4 to 1.6 cm long cytoplasmic fragments.
2. Nuclei were grafted onto RNase-treated cell fragments.
3. The nuclei were removed from the cell fragments after given time intervals, varying from 0 to 6 days.

4. Growth and protein synthesis in the enucleated fragments were measured.

It was found that if the nucleus remained attached to an RNase-treated cell fragment for 72 hours or more, the cytoplasm regained its capacity to proceed with protein synthesis and growth in the

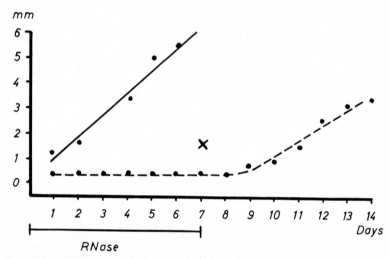

Figs. 18 and 20. Growth (Fig. 18) and protein synthesis (Fig. 20) of nucle-ated fragments 2 mm in length of *Acetabularia mediterranea* in a normal cul-ture medium (solid line) and in a RNase-containing medium for 7 days (dotted

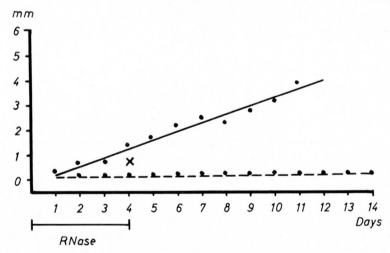

Figs. 19 and 21. Growth (Fig. 19) and protein synthesis (Fig. 21) of enucle-ated fragments 10 mm in length of *Acetabularia mediterranea* in a normal cul-ture medium (solid line) and in a RNase-containing medium for 4 days (dotted

absence of a nucleus. This experiment can be submitted to a serious criticism, since by grafting a normal nucleus to the cytoplasm, a variable amount of untreated cytoplasm is also transferred. Therefore it may not be justifiable to attribute the recovery of the treated cell fragments to the exclusive action of the grafted nucleus.

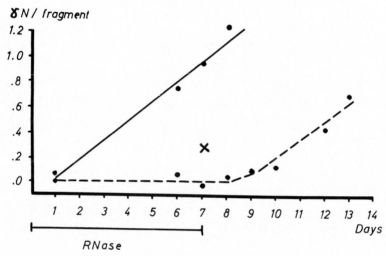

line). The fragments resume growth and protein synthesis after the transfer (X) to a normal medium.

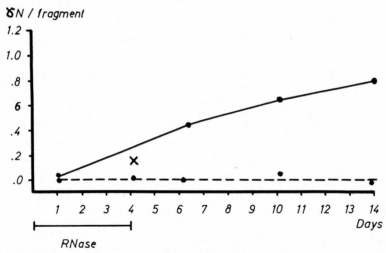

line). The fragments do not resume growth after a transfer (X) to a normal medium.

TABLE 9

Growth, Differentiation, and Protein Synthesis in *Acetabularia* Enucleated at Varying Periods After Treatment with RNase

Enucleation After RNase Treatment	Growth, mm	Differentiation	Protein Synthesis, γN
0 hours	0	0	0
24 hours	0	0	0
48 hours	3.6	1.7 whorls	0.42
72 hours	2.5	1.1 whorls	1.31
		1.2 caps	

To overcome this technical fault in the study just described, a new experiment was designed in which nucleated 1.5 cm long cells were subjected to an RNase treatment for 5 days. They were then transferred to a normal culture medium and after different time intervals (0 to 6 days) their nuclei were removed. Growth, differentiation, and the increase in amount of protein in the cytoplasm in these enucleated cell pieces were finally measured (Table 9).

Cytoplasm deprived of its nucleus immediately following the RNase treatment was unable to grow or synthesize protein. If, however, the nucleus and cytoplasm remained in contact for 2½ days, the cytoplasm showed the first signs of recovery, and after 4 days the cytoplasm resumed completely its independence from the presence of the nucleus as previously described.

The simplest interpretation of the results obtained would be to assume that RNase treatment markedly depleted or blocked the RNA content of the cytoplasm and that consequently this diminished or inhibited RNA content was the cause of the interruption of protein synthesis and growth. Following this it would seem that only the nucleus could provide the cytoplasm with the one or more substances necessary for inducing the new synthesis of RNA, which could be followed by protein synthesis, growth, and differentiation. It seems likely that the nuclear substance secreted into the cytoplasm might be RNA itself. The high exchange rate of P^{32} of the nucleolar RNA seems to support this assumption and indicate the high importance of the nucleolar substances (Stich and Hämmerling, 1953; Stich, 1956a; Hämmerling and Stich, 1956a, 1956b). The results of experiments on amoebae demonstrating a transmission of labeled RNA from the nucleus into the cytoplasm also favor this hypothesis (Goldstein and Plaut, 1955). However, it would be a hazardous oversimplification to consider RNA as the only possible substance able to transfer information from the chromosomes to the cytoplasm. The existence and the behavior of the RNA-free nucleolar substance attached to one specific locus of the giant chromosomes (see p. 106) show that other compounds, in this case perhaps a nonbasic protein, can also be considered.

Concluding Remarks

The experiments described have been restricted to two subjects, *Chironomus* and *Acetabularia*, and to two problems, namely, the relationship between chromosome loci and nucleolar substance and the interrelationships between the function of nucleoli and cytoplasmic activity. The results obtained appear to justify the conclusions that (1) substances of different chemical composition and different metabolic activity accumulate at different loci of the chromosome and on the other hand that (2) the amount of energy-rich phosphates in the cytoplasm regulates the volume, metabolism, and function of nucleoli. If these results can be generalized, then they can perhaps be used in speculation concerning the timing of gene action, which acquires special importance for normal growth. It may be that cytoplasmic factors regulate the activity of different chromosomal loci, resulting in the production of nucleolar substances with different compositions.

Finally, it was possible to demonstrate function of the interphase nucleus by grafting normal nuclei to RNase-treated cytoplasm. From present evidence it can be suggested that the nucleus contributes to the cytoplasm one or more compounds, which act as "starter molecules," and that in the cytoplasm an "amplifier effect" takes place. This assumption offers the simplest explanation for the dependence of the RNase-treated cytoplasm on the activity of the nucleus and on the other hand it also makes plausible the assumption that there is an autonomy of cytoplasmic RNA and protein synthesis once the nuclear substances have been released into the cytoplasm.

The results and interpretations mentioned above may also throw some light on the discussion (Mazia and Prescott, 1955; Hämmerling and Stich, 1956a; Brachet and Chantrenne, 1956; Plaut and

Rustad, 1957) caused by the differences in behavior of enucleated cytoplasmic fragments of *Amoeba* and *Acetabularia*. In the *Amoeba* enucleation suppresses the cytoplasmic net synthesis of RNA and proteins, whereas in *Acetabularia* the extirpation of the nucleus shows no immediate effect on cytoplasmic metabolism. That these differences are not fundamental in nature is shown by the RNase-treated cytoplasm of *Acetabularia,* which resembles the enucleated *Amoeba* in being strictly dependent on the presence of the nucleus. It seems that a difference exists only in the degree of cytoplasmic autonomy or, as previously described, in a different efficiency of the "amplifier effect." This can perhaps be explained by the fact that the energy production by *Acetabularia* continues unchanged because of the unaltered photosynthetic capacity of the plastids, whereas the enucleated amoebae fall into a state of starvation.

References

ALFERT, M., and T. GESCHWIND. 1953. A selective staining method for the basic proteins of cell nuclei. *Proc. Nat. Acad. Sci.* 39: 991–999.

BEERMANN, W. 1956. Nuclear differentiation and functional morphology of chromosomes. *Cold Spring Harbor Symp. Quant. Biol.* 21: 217–230.

BRACHET, J. 1953. The use of basic dyes and ribonuclease for cytochemical detection of ribonucleic acid. *Quart. J. Microsc. Sci.* 94: 1–10.

BRACHET, J. 1956. The mode of action of ribonuclease on living root tips. *Biochim. Biophys. Acta.* 19: 583.

BRACHET, J., and H. CHANTRENNE. 1951. Protein synthesis in nucleated and non-nucleated halves of *Acetabularia* studied with carbon-14 dioxide. *Nature.* 168: 950.

BRACHET, J., and H. CHANTRENNE. 1952. Incorporation de $C^{14}O$ dans les protéines des chloroplastes et des microsomes de fragments nucléés et enucléés d'*Acetabularia mediterranea*. *Arch. intern. Physiol.* 60: 547–549.

BRACHET, J., and H. CHANTRENNE. 1956. Function of the nucleus in the synthesis of cytoplasmic proteins. *Cold Spring Harbor Symp. Quant. Biol.* 21: 329–337.

BRACHET, J., H. CHANTRENNE, and F. VANDERHAEGHE. 1955. Recherches sur les interactions biochimiques entre le noyau et le cytoplasme des organismes unicellulaires. II. *Acetabularia mediterranea*. *Biochim. Biophys. Acta.* 18: 544–563.

CASPERSSON, T. *et al.* 1950. *Cell Growth and Cell Function.* W. W. Norton & Co., Inc., New York.

CHANTRENNE–VAN HALTEREN, M. B., and J. BRACHET. 1952. La respiration de fragments nucléés et enucléés d'*Acetabularia mediterranea*. *Arch. internat. Physiol.* 60: 187.

CLEMMONS, J. 1955. Procedures and errors in quantitative historadiography. *Biochim. et Biophys. Acta.* 17: 297–321.

CLEMMONS, J., J. LALICH, and D. ANGEVINE. 1954. The technique of quantitative historadiography. *Lab. Invest.* 3: 19–32.

ENGSTRÖM, A. 1946. Quantitative micro- and histochemical elementary analysis by Roentgen absorption spectrography. *Acta Radiol.* Suppl. 63. 106 pp.

ENGSTRÖM, A., and B. LINDSTRÖM. 1950. A method for the determination of the mass of extremely small biological objects. *Biochim. et Biophys. Acta.* 4: 351–373.

GOLDSTEIN, L., and W. PLAUT. 1955. Direct evidence for nuclear synthesis of cytoplasmic ribose nucleic acid. *Proc. Nat. Acad. Sci. 41:* 874–880.

HÄMMERLING, J. 1934. Entwicklungsphysiologische und genetische Grundlagen der Formbildung bei der Schirmalge *Acetabularia. Naturwiss. 22:* 829–836.

HÄMMERLING, J. 1943. Entwicklung und Regeneration von *Acetabularia crenulata. Z. induk. Abst. Vererbungslehre. 81:* 84–113.

HÄMMERLING, J. 1946. Neue Untersuchungen über die physiologischen und genetischen Grundlagen der Formbildung. *Naturwiss. 33:* 337–342, 362–365.

HÄMMERLING, J. 1953. Nucleo-cytoplasmic relationship in the development of *Acetabularia. Internat. Rev. Cytol. 2:* 475–498.

HÄMMERLING, J., and H. STICH. 1956a. Einbau und Ausbau von ^{32}P im Nucleolus (nebst Bemerkung über intra- und extranukleare Proteinsynthese). *Z. Naturf. 11b:* 158–161.

HÄMMERLING, J., and H. STICH. 1956b. Abhaengigkeit des ^{32}P in den Nucleolus vom Energiezustand des Cytoplasmas. *Z. Naturf. 11b:* 162.

JOHNSON, M. J. 1941. Isolation and properties of a pure yeast polypeptidase. *J. Biol. Chem. 137:* 575–586.

MAZIA, D., and D. PRESCOTT. 1955. The role of the nucleus in protein synthesis in *Amoeba. Biochim. Biophys. Acta. 17:* 23–34.

MONNÉ, L. 1950. The disappearance of protoplasmic acidophilia upon deamination. *Arch f. Zool. 1:* 455–462.

PELC, S., and A. HOWARD. 1952. Techniques of autoradiography and the application of the stripping film method to the problem of nuclear metabolism. *Brit. Med. Bull. 8:* 132–135.

PLAUT, W., and R. RUSTAD. 1957. Cytoplasmic incorporation of a ribonucleic acid precursor in *Amoeba proteus. J. Biophys. Biochem. Cytol. 3:* 625–630.

STICH, H. 1951a. Experimentelle karyologische und cytochemische Untersuchungen an *Acetabularia mediterranea. Z. Naturf. 6b:* 319–326.

STICH, H. 1951b. Trypaflavin und Ribonucleinsaure. Untersucht an Mäusegeweben, *Condylostoma* spec. und *Acetabularia mediterranea. Naturwiss. 18:* 436.

STICH, H. 1953. Der Nachweis und das Verhalten von Metaphosphaten in normalen, verdunkelten und Trypaflavin-behandelten Acetabularien. *Z. Naturf. 8b:* 36–44.

STICH, H. 1955. Synthese und Abbau der Polyphosphate von *Acetabularia* nach autoradiographischen Untersuchungen des ^{32}P-Stoffwechsels. *Z. Naturf. 10b:* 281–284.

STICH, H. 1956a. Bau und Funktion der Nukleolen. *Experientia. 12:* 7–14.

STICH, H. 1956b. Änderungen von Kern und Polyphosphaten in Abhängigkeit von dem Energiegehalt des Cytoplasmas bei *Acetabularia. Chromosoma. 7:* 693–707.

STICH, H., and J. HÄMMERLING. 1953. Der Einbau von ^{32}P in die Nucleolarsubstanz des Zellkernes von *Acetabularia mediterranea. Z. Naturf. 8b:* 329–333.

STICH, H., and A. KITIYAKARA. 1957. Self-regulation of protein synthesis in *Acetabularia. Science 126:* 1019–1020.

STICH, H., and W. PLAUT. 1958. The effect of ribonuclease on protein synthesis in nucleated and enucleated fragments of *Acetabularia. J. Biophys. Biochem. Cytol. 4:* 119–121.

STICH, H., S. M. GRELL, and J. McINTYRE. 1958. Studies on giant chromosomes and nucleoli of *Chironomus* using the autoradiographic and X-ray absorption techniques. (In press).

TAYLOR, J., R. D. McMASTER, and M. F. CALUYA. 1955. Autoradiographic study of incorporation of P^{32} into ribonucleic acid at the intracellular level. *Exp. Cell Res. 9:* 460–473.

THILO, E., H. GRUNZE, J. HÄMMERLING, and G. WERZ. 1956. Über Isolierung und Identifizierung der Polyphosphate aus *Acetabularia mediterranea. Z. Naturf. 11b:* 266–270.

VANDERHAEGHE, F. 1954. Les effets de l'énucléation sur la synthèse der protéines chez *Acetabularia mediterranea. Biochim. Biophys. Acta. 15:* 281–287.

VANDERHAEGHE, F., and D. SZAFARZ. 1955. Énucléation et synthèse d'acide ribonucléique chez *Acetabularia mediterranea. Arch. intern. Physiol. 63:* 267–268.

VINCENT, W. 1955. Structure and chemistry of nucleoli. *Internat. Rev. Cytol. 4:* 269–298.

WAGNER-JAUREGG, T. 1943. Die neueren biochemischen Erkenntnisse und Probleme der Chemotherapie. *Naturwiss. 31:* 335–344.

WERZ, G. 1955. Kernphysiologische Untersuchungen an *Acetabularia. Planta. 46:* 113–153.

7

Developmental Changes in Chloroplasts and Their Genetic Control

DITER VON WETTSTEIN [1]

Although a fairly accurate picture of the submicroscopic structure in mature chloroplasts had been presented some 20 years ago, it is only a few years since the electron microscope has enabled us to inquire into the details of this structure, its formation during plant ontogenesis, and its changes under different physiological conditions.

Interests in the cytology of chloroplasts center around three fields: (1) *general cell biology*, (2) *photosynthesis*, and (3) *genetics of chlorophyll deficiencies.*

1. The chloroplast is the most prominent inclusion of the cytoplasm in a plant cell and has a complicated but highly ordered layer structure. These relatively large layers of protein, lipoprotein, or chromolipoprotein are, at least in the higher plants, built up from unordered material in small organelles, the so-called proplastids. The analysis of this process may disclose a general mechanism for the morphogenesis of layered structures. Plastids in cormophytes as well as chromatophores of thallophytes are capable of division even in a differentiated stage. We are thus confronted with the problem of the multiplication of a multilayered cell organelle.

2. The biochemistry, biophysics, and physiology of photosynthesis has provided a wealth of information. What is the relation of

[1] Genetics Department, Forest Research Institute and Institute of Genetics, University of Stockholm, Sweden. This work was generously supported by grants from the Swedish Agricultural Research Council. I would also like to acknowledge the invaluable assistance of H. R. Jung and Miss U. Edén.

chloroplast structure to photosynthesis and where do the different part reactions take place? In the present stage this is primarily a question of cytochemistry on a macromolecular level, that is to say, the localization of pigments and enzymes, of proteins and lipids in the structures that may be seen in the electron microscope. Here too belong questions such as energy transfer across or along chloroplast lamellae, storage of starch, or transport of products from photosynthesis within or away from the plastids.

3. Chlorophyll deficiencies, lethal or viable, may be conditioned by mutations in the genome, in the plasmone, or in the plastome (sometimes the word plastidome is used as synonym). Gene mutations affecting the development and function of chloroplasts are excellent material for the study of gene action on the synthesis of submicroscopic structures. Plasmone- or plastome-conditioned chloroplast characters on the other hand may help us to understand the effect and nature of these two genetic systems outside the nucleus. Plastids are often ascribed far-reaching genetic autonomy. This rests essentially on two assumptions: first, that the plastids are exclusively autonomous self-duplicating organelles, that is to say, that they can multiply only by division. Second, that the plastome (the genetic system outside the nucleus determining certain plastid characters) is "localized" in the plastids. On both assumptions no general agreement has been reached so far.

I will restrict myself to certain topics related to the problems set forth here. (For a more comprehensive description of facts and for the older literature reference is made to Frey-Wyssling, 1953, 1955; Granick, 1955; Gustafsson and v. Wettstein, 1957; Leyon, 1956; v. Wettstein, 1957a, 1957b).

Chloroplast Structure in Different Species

The genetics of plastid characters in *Oenothera* has led Renner (1936) to the conclusion that the quality of the chloroplasts even in related species is fundamentally different. Quantitative structural differences as to size of the grana in the chloroplasts of different tissues, of plants in different physiological conditions, and of different species as well as genera are well known from light microscopic studies. What then are the similarities and dissimilarities of the submicroscopic chloroplast structures in different genera of thallophytes and cormophytes so far investigated?

The basic structure of the chromatophores in algae consists of continuous layers about 30 mμ thick. These layers are composed mostly of four lamellae with a thickness of 60 ± 20 Å. A chromatophore of the brown alga *Fucus vesiculosus* (Leyon and v. Wettstein, 1954) may serve as an example (Plate X–1). A schematic representation may be found in Fig. 22. The structural elements of the chromatophores throughout the thallophytes from flagellates to *Nitella* show these same structural elements, including even big and morphologically complicated chromatophore types such as in *Spirogyra* (for literature see Leyon, 1956; v. Wettstein, 1957a). So far none of the thallophytes investigated has revealed the type of grana differentiation characteristic of higher plants. Whereas in many algae the reported structures appear to consist of continuous lamellae throughout the whole chromatophore, the chromatophores of *Euglena* and certain desmids are claimed to contain separated stacks of individual discs (Sager and Palade, 1954, 1957; Chardard and Rouillier, 1957). It seems to me that this may be an expression of different physiological stages of the chloroplasts, especially since some of the pictures were obtained after cultivation of the organism in the dark for 3 weeks. The lamellae in some algae can be broken down and built up again from undifferentiated structures (Wolken and Palade, 1953) and this appears to happen in a similar way as that known from higher plants, i.e., short and closed double lamellae fuse to form large continuous lamellae. In a similar way the chromatophore may be reorganized locally under certain physiological conditions with the aid of such separated short lamellar discs. It is to be noticed too that in algae, just as in higher plants, free ends of lamellae never appear. They are always connected in pairs. Even the layers can be connected at the chromatophore ends to form continuous loops (cf. Plate X–1).

Regarding the submicroscopic structure of chloroplasts in higher plants, I would like to present that of barley in some detail. Plate VIII gives the cross section of a lens-shaped chloroplast in the mesophyll of a primary leaf 15 cm long. In the granular ground substance, the stroma, lamellated grana may be seen connected by the stroma lamellae. Especially in the upper half of the picture it can be seen that the grana and stroma lamellae are aggregated in layers transversing the entire chloroplast in a more or less regular way. The grana viewed perpendicular to the lamellae (face view of the plastid) are known to be circular. Thus the grana are cylinder stacks of specialized parts of the lamellae that are relatively opaque in the electron microscope. The chloroplast is surrounded by a twofold membrane; beneath the membrane is located the so-called peristroma. This peripheral region of the ground substance is generally not occupied by lamellae. Its peculiarity can be traced from certain *Oenothera* mutants and in slightly swollen plastids (Plate X–5). In addition to the mentioned structures a few round black globules can be seen. In referring to this plastid component we will simply use the term *globuli*.

Plate IX, 1–4, gives the details of the organization within the plastid. Plate IX–1 shows the general structural specialization of the chloroplast into grana and stroma. Plate IX–3 reveals the highly parallel arrangement of the grana lamellae, which measured from 40 to 60 Å thick. Sometimes each lamella can be resolved into two fine layers. The number of lamellae in the stroma and grana is the same. However the stroma lamellae are thinner (20–30 Å) and always closely paired two by two (Plate IX–2). The stroma lamellae display surface asymmetry, i.e., the space enclosed by the two paired lamellae appears empty, whereas the space between pairs of stroma lamellae is filled with the granular ground substance. In some micrographs (cf. Plate IX–1, 2, 4) an additional line is found in the middle of this space. It is not clear whether this is a real structure or some sort of artifact.[2] Of special interest is the connection between the grana and stroma lamellae. This can be studied in Plate IX–4. The righthand portion shows a granum, the left, a region of the stroma. Each of the paired stroma lamellae is continuous with one grana lamella. Just at the point of connection the lamellae are bent apart in such a way that the distance between the grana lamellae becomes equal. The space between the grana lamellae is about equal to the thickness of the lamellae. At the edge of the chloroplast and, in some cases in its center, the paired stroma lamellae are connected to form "loops" (Plate VIII and Plate IX–1, upper left corner). This can be demonstrated especially easily in slightly swollen chloroplasts (Plate X–4, 5). Here the stroma lamellae enclose an empty "white" space and separate it from the surrounding granular peristroma or stroma. We thus arrive at the schematic organization of the barley chloroplast in Fig. 22. In the same diagram is portrayed the effect of the slight swelling that sometimes occurs during fixation and

---

[2] Since this chapter went to press, it has been shown that these lines represent an artifact caused by refraction effects in layered structures.

embedding. The grana are rather resistant to swelling in comparison with the stroma lamellae. Interestingly enough, the swelling separates the paired stroma lamellae and leads to the impression of a reverse pairing. The pairing seems to be accomplished by the granular ground substance (Plate X–2) and the widened intralamellar space appears empty. From these facts it is concluded that

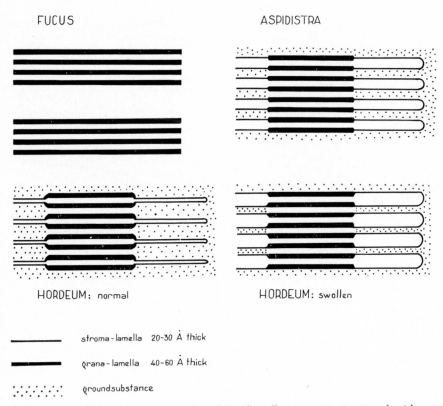

Fig. 22. Schematic representation of the lamellar organization in plastids.

there should be a phase boundary between the stroma and grana, since the latter do not display these effects. This is accentuated by the observation that the grana lamellae are generally connected at their border by some diffuse material, often giving the impression of grana discs. The surface asymmetry of the stroma lamellae and their closed connection at the end of the chloroplast allows the granular material of the peristroma to reach far into the interior of the plastid in flattened channels (Plate X–4, 5). It may be important to determine whether natural swelling gives us the same separation

of lamellae and whether a stage such as in Plate X–2 is reversible.
So far we can only imagine the implications of this structural organization in hypotheses: The surface asymmetry of the stroma lamellae
may have something to do with a separation in space of synthesis
and transport of low molecular substances as well as the storage
of products of photosynthesis. It is of interest that Calvin and his
co-workers (1955, 1956) assume such a surface asymmetry of chloroplast lamellae for the connection of the photochemical apparatus
with the carbon cycle and the Krebs cycle.

The structure of the chloroplasts of the other higher plants investigated so far fits well into the picture drawn here for the barley
chloroplast. These are *Aspidistra elatior* (Leyon, 1956; Steinmann
and Sjöstrand, 1955); *Chlorophytum comosum* (Strugger, 1957);
Zea mays (Hodge *et al.*, 1955, 1956); *Oenothera* (Stubbe and v.
Wettstein, 1955); *Solanum lycopersicum* (Lefort, 1957a, 1957b);
and *Picea excelsa* (v. Wettstein, unpublished). In all these different plants the thickness of the grana lamellae is about 60 Å and the
stroma lamellae measure about half this thickness. Although details
in interpretation vary, the reported pictures bear out the structural
organization described previously. In Plate X–3 the lamellar organization at the stroma-grana border for *Aspidistra* (Leyon, 1956) is
given. It is easily seen how closely it resembles the picture of a
slightly swollen barley chloroplast (Plate X–2 and Fig. 22). The
chloroplast of *Chlorophytum* has the same structure as that of
Aspidistra. In *Zea mays* the reported micrographs resemble those
of the barley chloroplast. We have reported some different structural organization for the *Oenothera* chloroplast. There was no
continuous lamellar system. Some grana lamellae formed individual
discs and grana consisting of 6, 8, or 10 lamellae were connected
only by 2 or 4 paired stroma lamellae. Some recent studies performed by M. Lefort in our laboratory on the tomato plant revealed
a similar arrangement in young chloroplasts, but older chloroplasts
had the typical 1:1 relation between stroma and grana lamellae.
This shows that grana formation and function of the chloroplast may
precede a continuous lamellar system. Probably the structure of the
Oenothera chloroplast represents a young stage that eventually will
develop into the typical structure.

Thus it appears that chloroplasts of widely different genera have
a very similar submicroscopic structure; and in Fig. 23 a general
diagram of the organization of the chloroplast in higher plants is

presented. The only serious difference revealed between different genera concerns the degree of pairing of the stroma lamellae, but this is probably nothing more than the reflection of different physiological stages. The diagram brings out another fact that has led to considerable discussion, namely, the position of the grana. Both in corn and barley it can be shown that in relatively young but fully active chloroplasts the layer structure is pronounced and that individual grana within these layers are not lying perpendicular

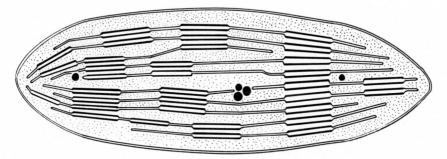

Fig. 23. Schematic picture of the submicroscopic chloroplast structure.

above one another (left part of the diagram). As the chloroplasts grow older the number of lamellae is increased and a pairing of the grana regions takes place, leading to grana columns traversing the plastid as a whole. Thus even the mature active chloroplast differentiates continuously. This ever-present reorganization is also reflected by some tracer studies showing that half of the chlorophyll content in a rye leaf is renewed within 24 hours (Godnev, 1955) or by the fact that the lamellar organization of chloroplasts may be changed into a fibrillar organization of the chromoplasts in the hips of roses (Steffen and Walter, 1955) and the fruits of *Capsicum* (Frey-Wyssling and Kreutzer, 1958a).

Differentiation and Division of Chromatophores in Algae

The so-called elementary chromatophores in the egg cell and in the vegetation point of *Fucus vesiculosus* have a structural organization that shows the essential characteristics of the full-grown chromatophore, i.e., layers composed of lamellae arranged like scales in an onion. However, there is a considerable difference in size of the plastid and in the number of layers (cf. Leyon and v.

Wettstein, 1954). A comparison of Plate XI–2 and Plate XI–3 representing chromatophores in egg cells with a chromatophore from an old cortex cell in the thallus (Plate XI–1) at the same magnification may demonstrate the difference. Whereas the elementary chromatophores contain from 8 to 10 layers, their number is increased to as much as 18 during growth and differentiation. The multiplication of the layers seems to involve a process of thickening and subsequent cleavage, as seen in Plate XI–2, 3. The separation of the newly formed layers is often incomplete at their terminal ends, thus leading to closed double layers (Plate XI–3). These observations suggest that the formation of new lamellae in the algal chromatophores can take place in close connection with existing ones.

Several examples are known in which layers and lamellae can be formed in small or large organelles lacking a lamellar or layered structure. This is the rule in *Nitella* (Hodge *et al.,* 1956), in which the lamellae are developed during ontogenesis from small undifferentiated organelles. Here the continuous lamellae are formed by fusion of small flattened vesicles in the same manner as in higher plants (cf. following discussion). They differ from the latter by the lack of grana. Some strains of *Euglena, Poteriochromonas,* and *Chlamydomonas* (Wolken and Palade, 1954; Sager, personal communication) contain a lamellar structure during the whole life cycle, but the chromatophore lamellae are almost completely lost in the dark. When such cultures are replaced in the light, the lamellae are readily re-formed.

The examples quoted reveal that layers and lamellae in algal chromatophores can be formed either in close connection with existing ones or from undifferentiated material. The synthesis of lamellae from undifferentiated material can take place also in algae, in which layered plastids are normally present during the whole life cycle. From present cytological information the lamellae cannot be assigned an autonomous genetic continuity.

Observations with the light microscope have shown that the chromatophores in many algae multiply by fission and that the division of existing chromatophores is the predominant mechanism by which chromatophores are handed over from one generation to the next. Fig. 24 gives the principles for the division of a layered lens-shaped chromatophore in *Fucus vesiculosus* as found in young cortical cells and in Plate XI, 4–6 some of the stages are reproduced

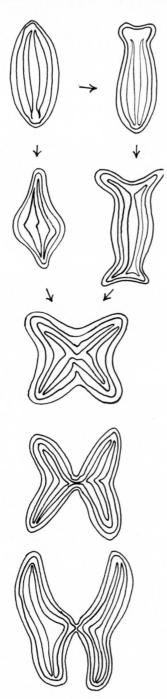

Fig. 24. Schematic sequence of the division of a layered chromatophore in *Fucus*.

from micrographs (cf. v. Wettstein, 1954). The chromatophore forms pseudopodlike protrusions and successive constrictions running parallel through all layers. Thereby the layers are brought into close contact. First the innermost two layers are combined and pinched off in a plane vertical to their surface. By this process, separated but continuous layers are formed. When the two innermost layers are separated, the next pair undergo the same process and so forth until all layers are separated. Plate XI–6 gives one of the later stages in the division of a chromatophore into two daughter chromatophores. In this instance the peripheral layers are common to both daughter chromatophores, whereas the inner layers have separated.

Some observations suggest that one chromatophore may give rise simultaneously to three or four daughter chromatophores. Free lamellar ends do not arise during division. Another consequence of this type of fission of a chromatophore is that one end of the daughter chromatophore may contain concentrically arranged lamellae and the other end the parallel lamellae of the original chromatophore. In the further differentiation of the chromatophore the concentric arrangement of the lamellae changes and the parallel arrangement of closed, paired lamellae is re-established. It is also worth mentioning that after division the daughter chromatophores do not always have the same number of lamellae. This is because some lamellae may not be involved during fission.

Sometimes the separation of dividing chromatophores is incomplete even in *Fucus*. This leads to chains of two or three lens-shaped chromatophores or starlike bodies. We may assume that morphologically complicated chromatophores of certain algae as they are found, for instance, in some *Ectocarpales* or in *Zygnema* arise in this way.

These observations on the division of layered algal chromatophores support the general cytological theory of self-duplication of plastids. One of the more important arguments in this theory is an experiment with a particular strain of *Euglena* (Lwoff and Dusi, 1935). The cells of this strain lose their chromatophores by prolonged cultivation in the dark, and cannot regenerate them when replaced in light. It was the inhibition of chromatophore division by darkness that produced the loss of the plastids. Obviously this strain could not form chromatophores from cytoplasmic structures.

In the case of another strain of *Euglena*, which lacks the ability to form chlorophyll and to maintain its layer structure in the dark, Wolken and Palade (1953) report: "The chromatophores seem to vanish beyond recognition." In a later study (Wolken, 1956) the disintegration of the chloroplasts to "ghost" structures on dark adaptation was confirmed. Such cells put back into light could restore their chromatophores and lamellae.

This difference—chlorophyll formation in the dark and in the light versus chlorophyll synthesis only in light—can be brought about by mutation (Granick, 1951; Sager and Palade, 1954; Sager, 1955). The ability to maintain the lamellae in dark culture may also be lost by mutation. From these facts the following picture can be inferred: originally the algae had the ability to build up the layered chromatophore from undifferentiated organelles, possibly from cytoplasmic structures. For this a series of gene-controlled reactions is necessary, just as in higher plants. Then they acquired the mechanism for division of layered chromatophores, which again requires another series of gene-controlled reactions. Thirdly, they gained gene-controlled reactions that maintain the layered chromatophore in the dark or under other unfavorable conditions. During further evolution in the group of algae the genes for the formation of a layered chromatophore from undifferentiated structures may be lost or changed to other functions. As a result the organism depends entirely on the mechanism of division and the ability to maintain the chromatophore under all natural conditions. Such a picture is able to explain the observed facts. Even in the cases in which the chromatophores originate only from existing ones by division we need not assume autonomous hereditary factors in the layered chromatophore.

What happens when a single chromatophore in an alga that multiplies its chromatophores by division is modified in its structural organization? Van Wisselingh (1920) observed in *Spirogyra* a chromatophore without pyrenoids and therefore with abnormal starch formation together with normal chromatophores in the same vegetative cell. When the cell divided, both types divided and came to be present in the new cell. Probably this abnormality would have been carried over sexual reproduction and reproduced indefinitely as long as the chromatophores divided to give rise to new chromatophores. This is then exactly what we would recognize as autonomous plastid inheritance. Let us assume now an organism with an abnormal chromatophore, perpetuating this abnormality indefinitely by division. This organism, cultured in the dark, will lose its chromatophores, even the abnormal ones. However, chromatophores will be re-formed when the plant is replaced in light. According to our interpretation, the abnormality will not reappear, because the new chromatophore is formed from undifferentiated structures according to the genes present in the nucleus. This then would mean that during chromatophore division the existing structure determines the structure of the daughter chromatophores. It shows too that in algae, which have no ability to form chromatophores from undifferentiated material, the question of genetic factors in the plastids cannot be analyzed in this way, since the dark experiment is not possible.

Although chromatophores in algae are to a large extent self-duplicating organelles in a cytological sense, this need not necessarily mean that the chromatophores themselves or any of their components are genetically autoduplicants comparable with chromosomes and genes.

Development and Differentiation of Chloroplasts in Higher Plants

The schematic picture in Fig. 25 summarizes the views of the present author on the development of a proplastid to a chloroplast. It is based on the investigations of Leyon in *Aspidistra* and our own including *Hordeum, Solanum, Phaseolus,* and *Picea.*

In leaf meristems and vegetative points three types of organelles, which are suspected of being proplastids, are found. Plate XI–7 (a, b, c) shows these in a meristematic leaf cell of barley. One of them (a) contains strongly electron scattering, granular material. Leyon (1954) and Mühlethaler (1955) have proposed that such organelles are very young proplastids. The second type (c) is less electron scattering and contains a vesicular structure. Probably it will differentiate into mitochondria. In addition a relatively dense organelle enclosed by a double membrane with a structure very similar to the cytoplasm is found (b). It is either isodiametric or amoeboid in shape and often contains some vesicles. This type certainly constitutes a proplastid stage and its further development into chloroplasts can be traced. From electron-microscope studies so far

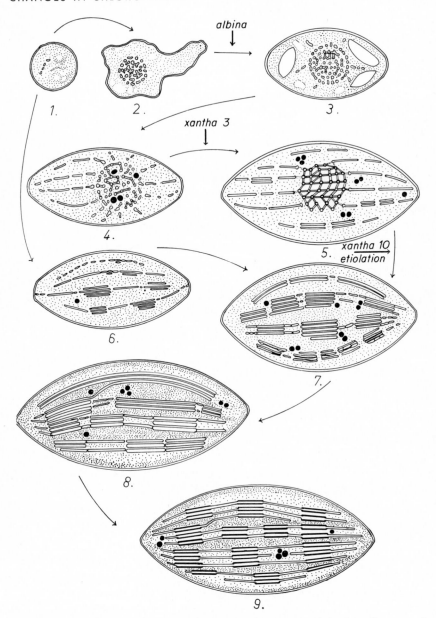

Fig. 25. Chloroplast development and morphogenesis of the lamellar organization in chloroplasts.

nothing definitely can be said about the origin of chloroplasts. Thus, the question whether they can only multiply by division or whether they can be condensed from the cytoplasm—just to mention two extremes—is still open.

The amoeboid-shaped proplastids seem to produce a considerable number of small vesicles, globular or flattened in shape (Plate XIV–1a). These display a strong tendency for aggregation and fusion. In barley it can be observed that they will either arrange themselves in longer or shorter chains than sometimes appear to be directly connected with the inner of the two enclosing membranes of the proplastid (Plate XIV–1a) or they will accumulate in the interior of the plastid to form plastid centers (core, prolamellar body, primary granum). One factor that is able to regulate this alternative is light. Barley seedlings grown under normal light conditions in the greenhouse generally do not reveal a typical center in the plastids of the apical or leaf meristems. As soon as a number of vesicles is found they aggregate in a few single rows and fuse to form four to eight parallel layers. Along a certain region of these layers the first short pieces of lamellae arise as groups of four to six parallel double discs (Plate XIV–1b). This is expressed in the schematic diagram of Fig. 25 by the arrow from Stage 1 to Stage 6.

If the germinating seedlings are kept under low light intensity or in complete darkness, large numbers of vesicles accumulate in one or several centers (Stage 2 of Fig. 25). In Plate XII–1 a somewhat older plastid of a barley seedling, germinated and fixed in complete darkness, contains two plastid centers, one cut in the median plane and the other obliquely (at the end of the plastid), as revealed from serial sections of the same plastid. In the centers a great number of vesicles, partly fused in different ways, can be seen. We have repeated this dark experiment three times with the same result.

Rapidly growing plants of tomato in light cultures possess in their shoot apices amoeboid proplastids with an inner differentiation corresponding to Stage 6 in Fig. 25 (Lefort, 1957a, 1957b). This is all the more remarkable since these belong to the youngest cells investigated in our laboratory and did not contain any vacuoles. None of the plastids investigated revealed a center. Similar results are found in the root meristem of *Zea mays* and *Vicia faba* (Heitz, 1957a). Plastids of Stage 6 were also found by Heitz (1957b) in an embryonal cell of the liverwort *Aneura*.

On the other hand, the aggregation of vesicles in a center has been demonstrated for etiolated corn (Hodge *et al.*, 1956). The center was also demonstrated in the root meristem of *Allium* (Strugger, 1957).

It looks as if these divergences can be explained by the findings in barley. Certainly other factors than light will be found to determine whether or not a center is formed during chloroplast development. Thus, if the formation of the first double layers and lamellae by fusion of vesicles can proceed at the same rate as the synthesis of the vesicles, no centers will appear. A delayed formation of layers and lamellae results in proplastids with one or several plastid centers. The physiological stage of the cell, plastid activity, the environment, and the genotype may determine the presence or absence of a center during plastid development.

Proplastids in the amoeboid stage seem to be capable of fission irrespective of their inner differentiation. In young proplastids starch grains are often accumulated (Plate XIV–1b).

If a plastid center is formed, the further development proceeds according to Fig. 25, Stages 3–5. Out of the plastid center vesicles protrude radially. They fuse to form larger and flattened discs. At the same time they arrange themselves in four to eight single layers. In the dark this process seems to be extremely slow, continuing for a long time, so that it is easy to catch. In the micrograph of Plate XII–1 and at a higher magnification (Plate XIII–2) a dark-grown barley plastid in this stage of differentiation can be traced. It is also to be seen from these micrographs that the vesicles and the larger discs are empty. This corresponds to the surface asymmetry found in the stroma lamellae in the full-grown plastid. In Plate XIII–1 a section through a plastid in a corresponding stage is shown, but here the section has been cut in such a way that a center was not hit. In this part of the plastid five continuous layers composed of single discs are completed. In a few places one layer consists of two flattened discs, which may be interpreted as a beginning of lamella formation.

It should be mentioned that in dark-grown barley the center may turn into a crystalline structure (Plate XIV–2). The crystalline appearance of the center has been found in *Aspidistra* and *Chlorophytum* even in light cultures (Leyon, 1954; Heitz, 1954[3]) and was

[3] Two days dark before fixation!

especially studied by Perner (1956). This crystalline structure will be discussed in connection with the process of etiolation below. In further development the crystalline center is incorporated into the lamellar organization and disappears.

The further development of the plastid is characterized by the formation of small pieces of lamellae within or in connection with the layers. This is a local process and takes place on mutually independent points.

At this point (Stage 7, Fig. 25) the two modes of plastid development, as concerns the formation of the layer and lamellar structure, meet. There is a considerable difference in over-all growth of the plastid in the two modes of differentiation. In the case of the formation of a center the main growth of the plastid takes place before and during the differentiation of the layer structure. This leads immediately to the formation of 8 to 10 layers. In this case the growth of the plastid precedes lamellar differentiation and the latter takes place when the plastid already has acquired its typical lens-shaped form.

If the layers and lamellae are built up without a center, first two, three, or four layers and a few lamellar discs appear. The plastids become lens-shaped after this differentiation. Simultaneously with the plastid growth, new layers are successively formed and the old layers enlarge in surface.

The subsequent steps during plastid development consist of a multiplication of the lamellar discs (Stage 7, Fig. 25), followed by a surface growth of the discs and their fusion to form continuous lamellae throughout the plastid (Stage 8). Then the differentiation of grana takes place and thereby the chloroplast structure is completed (Stage 9).

The multiplication of the short lamellar discs occurs in the proximity of the already existing layers. New discs may arise in two ways: by the fusion of small flattened vesicles that aggregate at certain points of the layers or by a process of thickening and fission of the primary lamellar discs in the layer, in analogy to the process found in *Fucus* (cf. previous discussion). Hodge *et al.* (1956) favor the first alternative. In barley we have found evidence for the second alternative (v. Wettstein, 1957a, 1957b).

One of the evidences consists of micrographs like Plate XIV–3. This is a portion of a young, developing barley plastid. The short pieces of lamellae show connections in different ways. They are very uneven in thickness and reveal splits of different orders of magnitude.

The newly formed lamellar discs come to lie closely parallel in stacks (Plate XIV–5). This picture, taken from a tomato plastid, reveals that the lamellar discs are very closely paired and the individual lamellae of two closely attached discs are often difficult to resolve. This gives the impression that the stacks consist of thick laminae bordered against the ground substance by a thin layer on both sides. This is especially remarkable in cases in which only two discs are paired (Plates XII–2, XIV–4), resulting in the picture of a thick line with two thin lines on both sides. However, the resolution of the thick lines into two lamellae of about 40 Å thickness and the connection of the lamellae at the ends of the stacks proves the stacks to be composed of equivalent lamellar discs (Plates XII–2, XIV–4, 5).

Obviously the lamellar discs grow in the direction of their plane until they come into contact with one another and fuse to form long continuous lamellae (cf. Plate XII–2). Even then no free ends of lamellae appear; they are always connected in pairs at their ends. For *Aspidistra*, barley, and corn, plastid stages with a lamellation throughout and without the grana structure have been found (Leyon, 1956; v. Wettstein, 1957b; Hodge *et al.*, 1955).

So far little is known about the differentiation of the grana structure. We do not know the exact relationships between the type of lamellae found in young plastids (Plates XII–2, XIV–4, 5) and the stroma and grana lamellae in a full-grown chloroplast (Plates VIII, IX–1).

Whereas in barley and *Aspidistra* the formation of a continuous lamellar system usually seems to precede the formation of grana, it looks as if in *Oenothera* (Stubbe and v. Wettstein, 1955) and in tomato (Lefort, 1957a, 1957b) a sort of grana differentiation occurs in the stage with the individual stacks of lamellae (cf. Plate XIV–5). This stage can be found in cotyledons and leaves of considerable size, which are certainly able to assimilate. However, in still older leaves, at least in the tomato, this organization turns into the typical chloroplast structure with the one to one relationship beween stroma and grana lamellae. In that case some of the stroma lamellae are secondarily built up between the grana. The globuli are also the main constituents in chromoplasts of the yellow petals in *Ranunculus* (Frey-Wyssling and Kreutzer, 1958b).

The globuli are present in clusters during development (see Stages 4 to 9, Fig. 25). They appear first within or close to the plastid center if a center has been formed. They have, as shown from a mutant analysis, a certain not yet understood function in the formation of the lamellae and grana structure as precursor, reserve, or waste material. They contain pigments (chlorophyll and/or carotinoids) and are probably chromolipids.

The picture of chloroplast development is far from being complete in all details. Generally the complicated lamellate structure of the chloroplast can be built up from structural elements not notably different from those found in the ground cytoplasm. According to the genetic constitution of the plant and its environment, the basic processes that lead to the submicroscopic organization are more or less modified. The basic processes are: (1) the synthesis of vesicles, (2) the fusion to flattened discs, (3) the arrangement of these discs in parallel layers, (4) the multiplication of lamellar discs, (5) growth and fusion of the lamellar discs to form a continuous lamellar system, and (6) the differentiation into grana and stroma regions. During the whole development no free ends of lamellae can be observed. The structural synthesis of the lamellae by fusion of vesicles to discs and from these to larger units provides a mechanism to produce surface asymmetrical lamellae as found especially in the stroma lamellae of a full-grown chloroplast.

Plastid Development in Etiolated Plants

Most angiosperm plants are unable to synthesize the chlorophyll pigments in the dark. They accumulate, however, protochlorophyll in the dark; this is readily converted into chlorophyll-*a* by photochemical transformation when the etiolated plant is subjected to light (Koski and Smith, 1948; Virgin, 1955a, 1955b). Smith and Kupke (1956) have recently isolated a protochlorophyll holochrome from etiolated bean seedlings that is able to carry on the photochemical transformation. They showed protochlorophyll to be associated with protein. Shibata (1957) in addition could show that the conversion of protochlorophyll to chlorophyll-*a* involves several steps. Our own studies in barley indicate that chlorophyll formation during germination in light proceeds in a different manner than chlorophyll formation during greening of etiolated plants (v. Wettstein, 1957a).

Therefore the development of chloroplast structures in etiolated plants in comparison to light-grown seedlings is of special interest. It might eventually give us some idea about the localization of the precursors to chlorophyll so successfully studied by spectrophotometry.

The first electron-microscope study of plastid development in the dark was performed by Leyon (1953) on young shoots of *Taraxacum*. He found that the plastids can develop to a considerable size, that plastid centers are formed, and a layer structure can be built up in the plastids. His most advanced stages show an organization that is comparable to Plate XIII–1 in the present paper.

Corn plastids show essentially the same development as that reported for *Taraxacum* (Hodge *et al.*, 1956). After two weeks of germination and growth in the dark these workers found plastids containing large centers consisting of minute vesicles as previously described.

It was stated in the investigation by Hodge *et al.* that lamellae did not show up in etiolated corn plants. Lamellae were supposed to be formed only when the plants are exposed to light. For barley at least this is not correct. Ten-day-old seedlings, measuring 10 to 15 cm in height, grown and fixed in absolute darkness, contain plastids with several centers and a layer structure, as seen in Plates XII and XIII. The layers consist of relatively large, single flattened discs. Thus in these etiolated plants the plastid development may be considered normal up to Stage 5 of Fig. 25. The mode of plastid development is, however, that characteristic for low light conditions.

If samples of etiolated leaves are taken after 22 days of growth when the primary leaves have reached a length of 25 cm, they exhibit a strong yellow color and are heavily curled. Under these conditions quite a different organization is revealed. If plastid centers are still present, they reveal a typical crystalline structure (Plate XVI) with a plane spacing of about 25 to 30 mμ, as measured from the micrographs. As seen especially in Plate XV–2, the crystalline structure is formed by the fusion of the original vesicles to one connected tubular system. (The plastid of this picture is somewhat swollen.) The tubular system is arranged in a very high degree of order, thus giving the crystalline appearance in sections. The "planes" of the lattice often consist of two dark lines enclosing a light space (Plate XVI), just as we would expect from a longitudinal section through a tube. The total diameter of the tube is about 60 to 90 Å and the diameter of the opaque tube wall measures 20 to 30 Å. The lattice "points" represent rings and are to be regarded as cross sections of the tubes. From a chemical point of view this crystalline structure must be rather complex.

From the crystallinelike structure the formation of layers can proceed even in the dark (Plates XV, XVI). However, layers do not arrange themselves normally in a parallel side-by-side fashion: a very regular concentric organization is accomplished. It consists of a system of circular, completely closed lamellae with an interspace of 25 mμ. This is the same value as found for the plane spacing in the crystal. In cross sections this leads to concentric rings of three to eight layers (Plate XV). From serial sections it can be seen that in one plastid several independent systems of concentric lamellae may be present.

The individual layers in this concentric system are composed of two opaque lines with a thickness of 20 to 30 Å enclosing a space of the same dimension (Plates XV, XVI). They thus have the same dimensions as the walls of the tubular elements in the crystal and it seems reasonable that the tubular elements of the "crystal" are transformed by a process of fusion into large double discs of globular form. In Plates XV–2 and XVI it may be seen in several places that the concentric double layers are directly connected with the tubular elements of the crystalline center. In addition it appears that the "planes" of the lattice, especially in the outer part of the plastid center, constitute larger pieces of double discs.

It remains to be seen whether these concentric layer systems are protruded out from the plastid center, as is suggested by pictures like that of Plate XV–2, or whether each plastid center forms a single concentric system of lamellae around itself (cf. Plate XVI).

The globuli, probably containing pigments, are always present in some clusters in these etiolated stages.

It is known that the early stages of etiolated plastids, such as in Plates XII–1 and XIII, can develop rapidly into chloroplasts when put into light. It is at present being investigated whether plastids with this concentrically arranged lamellar system can revert into normally structured chloroplasts or if this is an irreversible differentiation. We do not know so far what happens to the very first

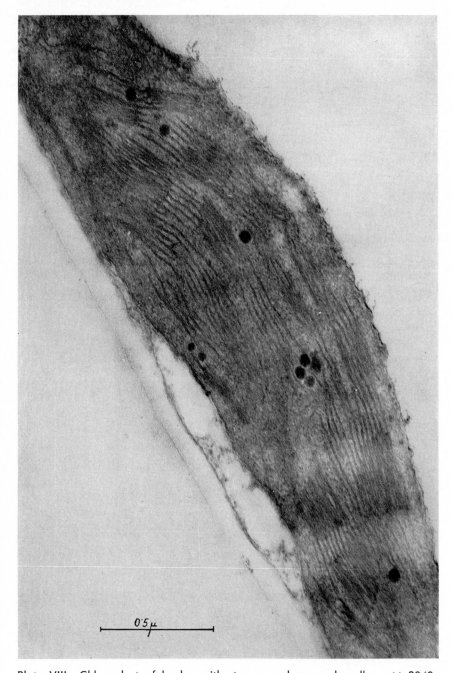

Plate VIII. Chloroplast of barley with stroma and grana lamellae. $\times$ 3840.

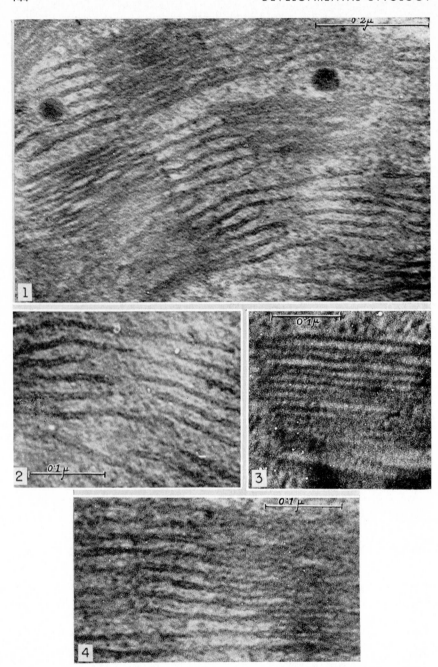

Plate IX. (1) Lamellar organization in grana and stroma regions. × 91,140 (2) Stroma lamellae. × 136,400. (3) Grana lamellae. × 136,400. (4) Connection between stroma lamellae (left) and grana lamellae (right). × 136,400.

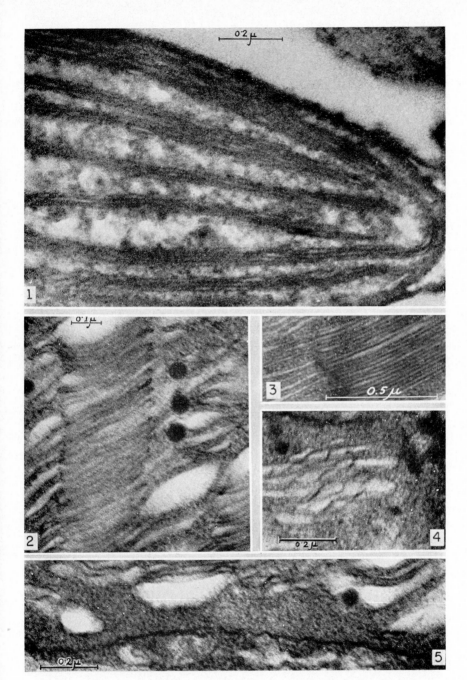

Plate X. (1) Chromatophore in the egg cell of *Fucus* with layers composed of lamellae. × 54,600. (2) Grana and stroma lamellae in a slightly swollen chloroplast of barley. × 49,800. (3) Stroma lamellae (left) and grana lamellae (right) of *Aspidistra* (Courtesy, Leyon, 1956). × 40,200. (4) and (5) Connection of the peristromium with the material between closed stroma double lamellae in a swollen chloroplast of barley. × 49,800.

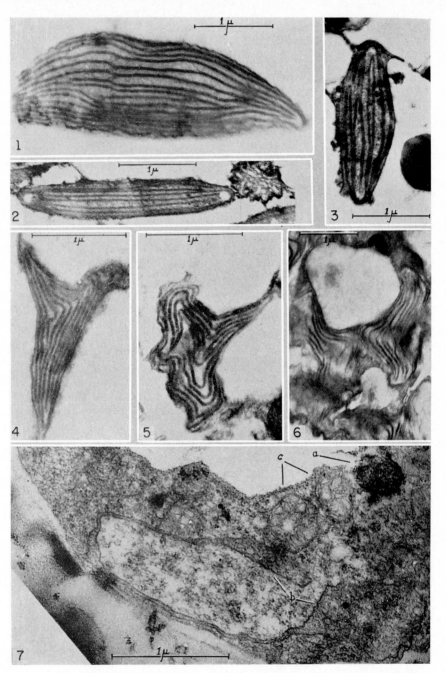

Plate XI. (1) Chromatophore of *Fucus* in an old cortical cell of the thallus. × 13,800. (2) and (3) Chromatophores from egg cells. × 13,800. (4), (5), and (6) Chromatophore division in *Fucus*. × 16,800 and × 10,200. (7) Meristematic leaf cell of barley with three different organelles (a, b, c). × 20,400.

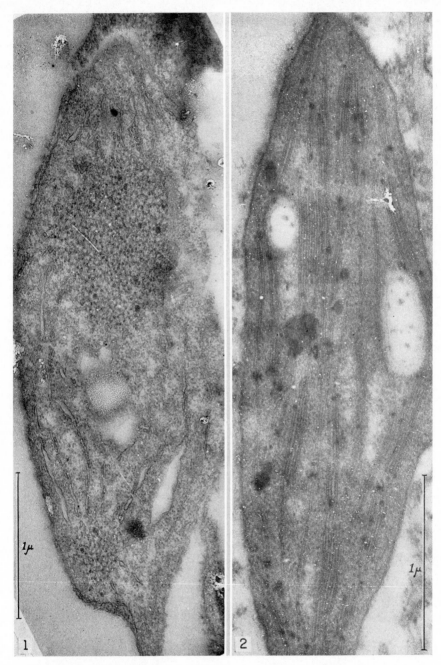

Plate XII. (1) Dark-grown plastid in barley with two centers. From the centers vesicles protrude and arrange themselves in layers. × 25,200. (2) Young plastid developing a continuous lamellar structure in the cotyledon of tomato. × 30,000.

147

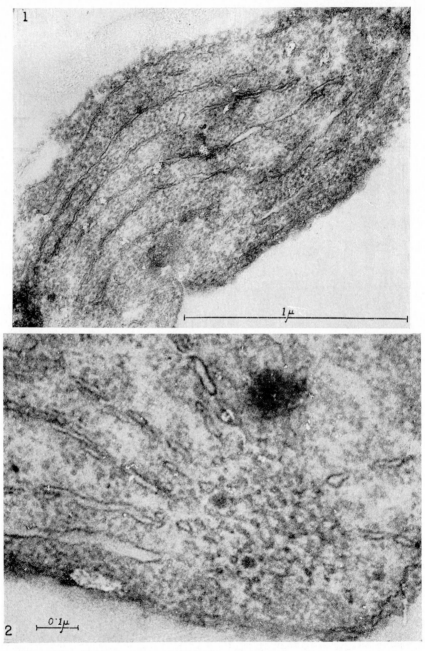

Plate XIII. (1) Dark-grown plastid of barley with the primary layer structure. × 40,200. (2) Part of a barley plastid. From the plastid center vesicles protrude in different directions. They fuse to form larger discs that arrange themselves in single layers. × 73,200.

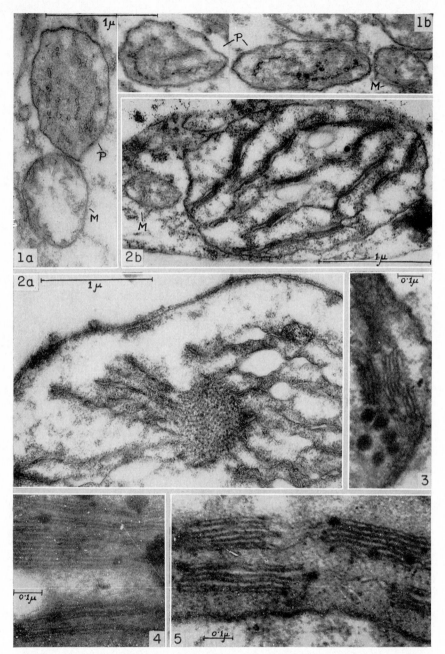

Plate XIV. (1) a. Young proplastids in leaf meristem of light-grown barley (P). b. Later stage. M = mitochondria. × 19,800. (2) a and b. Plastid of barley with crystalline center and primary layers. × 19,800. (3) Developing lamellar discs in barley plastid. × 44,400. (4) and (5) Lamellar structure in young plastids of tomato cotyledons. × 49,800.

149

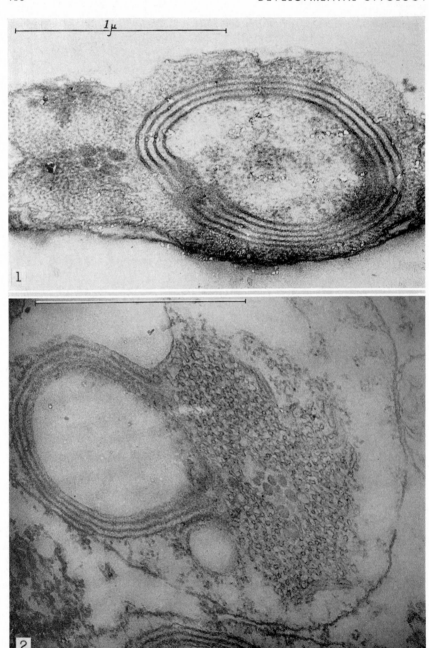

Plate XV. (1) Plastid of a three-week-old etiolated barley leaf, exhibiting concentric arrangement of lamellae. × 37,200. (2) Etiolated plastid with center and concentric layer structure. The center contains a tubular structure. × 37,200.

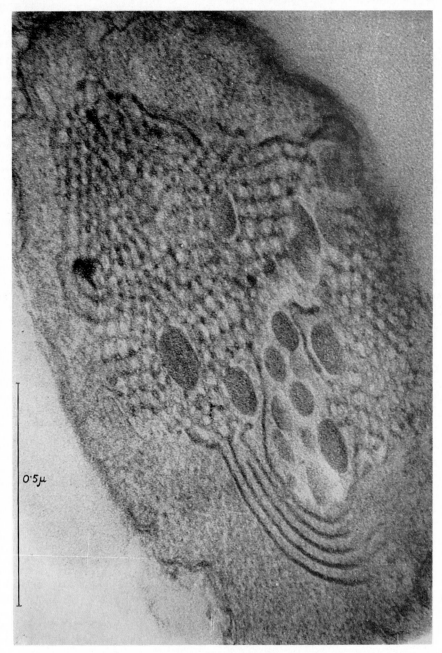

Plate XVI. Etiolated plastid with a crystalline center and some concentric layers. × 75,000.

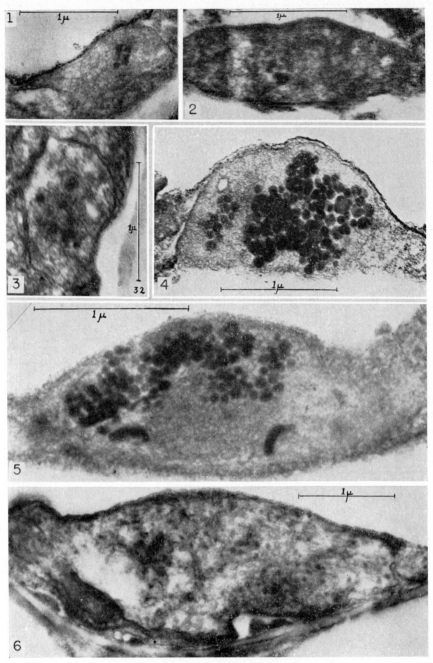

Plate XVII. (1), (2), (3) Different plastids in the gene mutant *albina*-20 of barley. × 16,800 and × 20,400. (4), (5), and (6) Stages of plastid differentiation of the gene mutant *xantha*-3 in barley. × 20,400, × 27,000, and × 16,800.

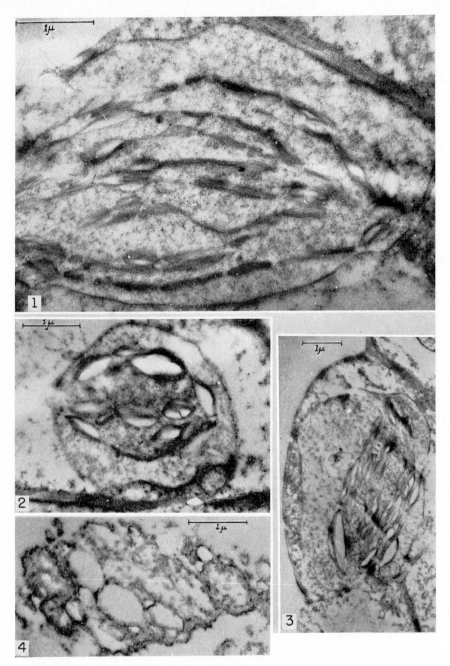

Plate XVIII. (1), (2), and (3) Chloroplasts of a yellow-green mutant in the *suaveolens* plastome of *Oenothera* (genome combination: *gaudens·*[h]*purpurata*). × 13,800, × 10,200, and × 6900. (4) Vacuolized chloroplast of a white mutant in the *suaveolens* plastome (genome: *Hookeri*). × 10,560. After Stubbe and v. Wettstein, 1955.

layers that appear in etiolated plastids of barley, whether these are incorporated into the concentric lamellar system or are broken down again.

We could conclude from our earlier studies (v. Wettstein, 1957a, 1957b) that for the primary synthesis of chlorophyll a lamellar structure is not necessary. From the results put forward here we can conclude that a lamellar chloroplast structure can be formed without chlorophyll. Is protochlorophyll holochrome necessary for this? In this connection the mutant *xantha*-10 is helpful. In this lethal mutant plastid development in the light leads to a structural organization with concentric lamellae (v. Wettstein, unpublished). In etiolated plants darkness prevents the normal arrangement of chloroplast lamellae, possibly as a direct consequence of the absence of chlorophyll. Etiolation constitutes from a structural point of view a phenocopy of *xantha*-10. The biochemical analysis of this mutant block revealed that the abnormal concentric arrangement of lamellae in the plastids of the mutant is correlated with a defect in the synthesis of protochlorophyll.

The Genetic Control

In an organism like barley several hundred genes control the development and structural differentiation of plastids. The invalidation of only one of these loci by mutation or deficiency may produce lethality. The mutations or deficiencies cause in homozygous condition white (*albina*), yellow (*xantha*), and light green (*viridis*) seedlings as well as types with different color patterns. These generally die as soon as the reserve substances of the seeds are exhausted. Only some light green mutants and variegated seedlings with sufficient green tissue can survive. Mutants with a darker green color and more chlorophyll than the mother strain can also be obtained.

As indicated in Fig. 25, such gene mutations causing chlorophyll lethals often imply blocks in the development of the submicroscopic plastid structure at certain points (v. Wettstein, 1954, 1957a, 1957b). Thus in *albina*-20 the differentiation is blocked between developmental Stage 2 and 3 of Fig. 25. In Plate XVII–1 and XVII–3 some micrographs of this mutant are reproduced. They show two plastids of a young secondary leaf. Some sort of plastid center can be produced as well as some globuli. The synthesis of layers and lamellae is prevented by the gene change. The plastid as a whole, however, continues to grow, nearly reaching the size of a mature chloroplast in spite of the lack of differentiation in the inner structure (Plate XVII–2). Thus the growth of the entire chloroplast is rather independent of its inner submicroscopic organization.

The mutant factor *xantha*-3 gives rise to a block in a later stage of the developmental chain (Fig. 25). It produces the plastid center and forms some primary layer structure (v. Wettstein, 1957a). The plastid differentiation cannot go on to the production of lamellae. Instead large numbers of globuli accumulate near the plastid center (Plate XVII–5). This production of globuli continues for a while until nearly the entire plastid is filled with globuli (Plate XVII–4). These stages are found in leaves 2 mm in length, and it seems reasonable to assume that the globuli accumulate because their functioning is arrested as a consequence of the gene change. This chlorophyll lethal is of the same nature as the nutritional mutants studied so extensively in *Neurospora* and other microorganisms. We have here a demonstration of a gene-controlled synthesis on the submicroscopic level and of an accumulation of a "structural intermediate," a plastid element of macromolecular size.

During the further growth of the plastids in *xantha*-3 the accumulated globuli are broken down. In 10 cm primary leaves, the interior of the plastid ends in a completely unordered state (Plate XVII–6).

That this conception of a block in the developmental chain leading to the lamellar organization of chloroplasts is probably valid for many chlorophyll lethals caused by mutations in the genome, is supported by the analyses of other factor mutants. We have already discussed the findings in *xantha*-10, in which the development is arrested between Stages 5 and 7 of Fig. 25, and the result is an inner organization typical for an advanced stage of etiolation. Recent studies of some mutants in tomato (Lefort, 1957a, 1957b) also revealed blocks at different points in the developmental chain of the plastid structures.

The viable *viridis* types on the other hand will often develop a fully differentiated chloroplast structure, as shown in the case of a mutant in corn (Hodge, 1956).

A correlation of these submicroscopic analyses with a biochemical characterization of the block in a chlorophyll lethal may give us some information regarding the nature of the plastid structures we

can observe in the electron microscope. *Xantha*-3 may serve us here
as a model: in Fig. 26 the amounts of pigments in different developmental stages are shown. These figures were obtained by analyzing

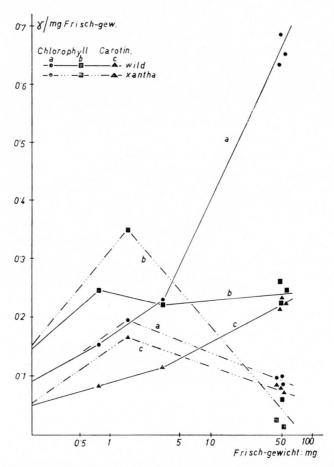

Fig. 26. Pigment content per mg fresh weight of different developmental
stages in normal barley seedlings (wild) and in the chlorophyll lethal *xantha-3*.

barley leaves of different sizes germinated and growing in light. In
the mother strain chlorophyll-*a* and carotinoids increase during leaf
growth and plastid development. Chlorophyll-*b* seems to be formed
at an early stage and its concentration seems to be constant
during the later plastid development. In a primary barley leaf 10
to 15 cm long, the normal proportions of chlorophyll-*a*, chlorophyll-
b, and carotinoids as well as the normal differentiated chloroplast

structure are established. The content of the pigments in *xantha*-3 during leaf development displays quite a different picture. The concentrations of chlorophyll-*a*, chlorophyll-*b*, and carotinoids follow optimum curves (broken lines). The optimum coincides with the plastid stage containing the greatest number of globuli (Plate XVII–4). It is significant that together with the accumulation of the globuli a larger amount of chlorophyll-*b* and carotinoids is found than in the case of normal plastids of a similar developmental stage. This strongly suggests that all the pigments are localized in the globuli. (For other evidence that the globuli contain pigments, especially carotinoids, see v. Wettstein, 1957a, and Frey-Wyssling and Kreutzer, 1958b.) With the breakdown of the globuli the pigments disappear. In a 10 cm leaf of *xantha*-3 with plastids of normal size but undifferentiated as to their inner structure (Plate XVII–6), the pigment content is greatly reduced in comparison with that of normal seedlings. Obviously chlorophyll synthesis is not blocked in this mutant; it can proceed in a quite normal manner. However, the pigments cannot be incorporated into a lamellar organization and therefore pigment synthesis stops. Thus the biochemical failure of this mutant is to be sought in the synthesis of another plastid component than chlorophyll.

In many cases it may not be possible to detect the accumulation of morphological or chemical precursors to a step which is blocked in a defective mutant. As soon as the rate of the degradation of the precursors approaches or reaches the rate of its production no accumulated intermediate is to be found. The rate of degradation is certainly influenced by outer and inner environmental factors. Thus the amount of pigments in 10 cm leaves of *xantha*-3 can be changed considerably by light intensity and temperature. This also holds for many other mutants (cf. Nybom, 1955).

Considering the different chlorophyll lethals caused by mutations in the genome, we can summarize the results in general: there are found genes that control the synthesis of plastid pigments. Here belong Granick's mutants in *Chlorella* (1954), which accumulate different porphyrins, or types in which the carotinoid synthesis is blocked (Claes, 1954). In higher plants such a case could be analyzed in *Arabidopsis* (Röbbelen, 1956). Here the photochemical and enzymatic conversion of protochlorophyll into chlorophyll-*a* is blocked, but the block can be avoided by the addition of an extract from etiolated normal plants.

There are certain genes that control the photostability of the newly synthesized pigments. Among others we have well-analyzed examples in *albina* mutants of corn, which can synthesize protochlorophyll in the dark in relatively large amounts. Upon illumination this is converted into chlorophyll-*a*. However, the latter is rapidly destroyed again (Koski and Smith, 1951). Since pigments constitute the plastid substances most easy to analyze, we are less well informed about genes that control the synthesis of proteins, nucleic acids, enzymes, etc., in the plastids. The findings in *xantha*-3, in which no block in pigment synthesis is found, indicate that even these other components are probably formed under genic control. Continued submicroscopic studies in combination with biochemical analyses will be required in order to show whether the gene-controlled failure of the differentiation of submicroscopic structures is always connected with the failure of the synthesis of a chemical compound or whether there are "true morphological" genes, as we may call them, which are specialized to control the arrangement of the compounds into a submicroscopic functional structure.

Regarding photosynthesis in its relation to plastid structure, those genes which exert their action on the plastids after the completion of a normal structure will constitute a valuable tool. All mutants that accumulate in their plastids structural elements of macro-molecular size, recognizable in the electron microscope, are of essential value for an indirect characterization of the structures in chemical terms. This can be accomplished by looking with chemical methods for those substances that are produced in amounts exceeding the normal level.

Chlorophyll lethals conditioned by mutation in the plastome have been analyzed with the electron microscope in *Oenothera* (Stubbe and v. Wettstein, 1955). So far the studies are restricted to a "yellow-green" and a "white" mutant of the *suaveolens* plastome. These mutants give rise to plastid defects in whatever genome combination they are placed and are inherited in a non-Mendelian fashion. Both types seem to reach an advanced submicroscopic organization with stroma and grana lamellae, thus presenting a structure comparable to normal green *Oenothera* chloroplasts.

When this organization has been reached the plastids begin to show signs of degeneration. In Plate XVIII–1 such an early stage of degeneration in the yellow-green mutant is reproduced. The chloroplast layers separate and the outer plastid layer, the peristromium, displays a sort of hypertrophy. Each plastid layer is composed of several lamellae and the individual grana can be seen. Obviously the grana are more resistant to the natural processes of swelling than the other plastid structures. Plate XVIII–2 and 3 show more advanced stages of degeneration. Now the vacuolization, well known from light-microscope studies, sets in and affects the grana as well. In these stages the widened peristromium can be recognized clearly, in addition to the plastid lamellae, which remain associated in a dense core. This degeneration process proceeds relatively slowly in the yellow-green mutant and can easily be followed. In addition it can be slowed down by low light intensities and temperature. In the white plastome mutant essentially the same processes occur; however, they proceed much faster and result in completely vacuolized plastids (Plate XVIII–4).

In contrast to the gene mutations just described, at least these two components of the plastome do not interfere with the development of the plastid structure. They start to work only when chloroplast development is completed. Further investigations will reveal if this distinction is a general one. It is supported by light-microscope findings in several other cases of plastid defects inherited in a non-Mendelian way (cf. v. Wettstein, 1957a, 1957b). Newly reported physiological investigations in *Oenothera* also point in the same direction (Schötz, 1955; Kandler and Schötz, 1956; Schötz and Reiche, 1957). We arrive for the moment at the hypothesis that the plastome does not control the development of the chloroplast. The genetic autonomy of the plastids, which is so strongly supported by the ingenious experiments of Renner in *Oenothera*, hereby becomes harder to understand. Nevertheless, the submicroscopic cytology of plastids together with the ever-increasing possibilities of biochemical analysis have provided us with tools for separating and defining the action of the genome, plasmone, and plastome on the plastids.

References

Comprehensive lists of literature have been published by several authors recently. We have chosen, therefore, to list these reviews in addition to a few articles not considered before.

CHARDARD, R., and C. ROUILLER. 1957. L'ultrastructure de trois algues Desmidées. *Extr. Rev. Cytol. Biol. Vég. 18*: 153–178.

FREY-WYSSLING, A. 1953. *Submicroscopic Morphology of Protoplasm.* Elsevier Publishing Co., Amsterdam. 255 pp.

FREY-WYSSLING, A. 1955. Die submikroskopische Struktur des Cytoplasmas. *Protoplasmologia,* II A, 2. 244 pp. Springer Verlag OHG, Vienna.

FREY-WYSSLING, A., and E. KREUTZER. 1958a. The submicroscopic development of chromoplasts in the fruit of *Capsicum annuum.* J. Ultrastruct. Res. *1:* 397–411.

FREY-WYSSLING, A., and E. KREUTZER. 1958b. Die submikroskopische Entwicklung der Chromoplasten in den Blüten von *Ranunculus repens.* Planta *51:* 104–114.

GRANICK, S. 1955. Plastid structure, development and inheritance. *Handbuch der Pflanzenphysiologie* I, 507–564. Springer Verlag OHG, Berlin-Göttingen-Heidelberg.

GUSTAFSSON, Å., and D. VON WETTSTEIN. 1957, Mutationen und Mutationszüchtung. *Handbuch der Pflanzenzüchtung* I, 612–699. Paul Parey, Berlin.

HEITZ, E. 1957a. Die Struktur der Chondriosomen und Plastiden im Wurzelmeristem von *Zea mais* und *Vicia faba. Z. Naturf. 12b:* 283–286.

HEITZ, E. 1957b. Die strukturellen Beziehungen zwischen pflanzlichen und tierischen Chondriosomen. *Z. Naturf. 12b:* 576–578.

LEFORT, M. 1957a. Structure fine du chloroplaste normal chez *Lycopersicum esculentum.* C. R. Acad. Sci. *244:* 2957–2959.

LEFORT, M. 1957b. Structure inframicroscopique des chloroplastes de certains mutants dépigmentés chez *Lycopersicum esculentum.* C. R. Acad. Sci. *245:* 718–720.

LEYON, H. 1956. The structure of chloroplasts. *Svensk kemisk tidskr. 68:* 70–89.

RENNER, O. 1936. Zur Kenntnis der nicht mendelnden Buntheit der Laubblätter. *Flora N.F. 30:* 218–290.

SAGER, R., and G. E. PALADE. 1957. Structure and development of the chloroplast in *Chlamydomonas. J. Biophys. Biochem. Cytol. 3:* 463–488.

SCHÖTZ, F., and G. A. REICHE. 1957. Untersuchungen an panaschierten Oenotheren II. *Z. Naturf. 12b:* 757–764.

SHIBATA, K. 1957. Spectroscopic studies on chlorophyll formation in intact leaves. *J. Biochem. 44:* 147–173.

SMITH, J. H. C., and D. W. KUPKE. 1956. Some properties of extracted protochlorophyll holochrome. *Nature 178:* 751–752.

STRUGGER, S. 1957. Elektronenmikroskopische Beobachtungen über die Teilung der Proplastiden im Urmeristem der Wurzelspitze von *Allium cepa. Z. Naturf. 12b:* 280–283.

WETTSTEIN, D. VON. 1957a. Chlorophyll Letale und der submikroskopische Formwechsel der Plastiden. *Exp. Cell Res. 12:* 427–506.

WETTSTEIN, D. VON. 1957b. Genetics and the submicroscopic cytology of plastids. *Hereditas 43:* 303–317.

WOLKEN, J. 1956. A molecular morphology of *Euglena gracilis* var. *bacillaris.* J. Protozool. *3:* 211–221.

8

Changes in the Fine Structure of the Cytoplasmic Organelles During Differentiation

Don W. Fawcett [1]

Until recently the cytologist interested in cell differentiation was largely limited to consideration of the changes in form, location, and number of those organelles that could be selectively stained. Since the development of satisfactory methods for the study of tissues with the electron microscope the scope of cytological investigations has been greatly extended. The submicroscopic structure of mitochondria, Golgi apparatus, and centrioles has been described, and a new cytoplasmic organelle, the endoplasmic reticulum, has been defined. Interest is now beginning to turn from purely descriptive studies to the analysis of changes in the fine structure of these organelles in successive stages of differentiation and in various physiological conditions. Owing to the restricted field of view in electron microscopy and the difficulty of consistent sampling, the choice of material is of considerable importance. The study of developmental changes in cells is greatly facilitated by working with an organ or organism in which multiple stages of differentiation of the same cell type are concentrated in a relatively small area. In this respect, the seminiferous tubules of the testis are ideal. Furthermore, the germinal epithelium has engaged the interest of many of the most able cytologists of the past, and the heritage of detailed and accurate observations they have left provides an excellent foundation for electron-microscope studies on the role of the cytoplasmic organelles

[1] Cornell University Medical College, New York. The work described was supported by Grant C-2623 from the National Institutes of Health and Grant E-16 from the American Cancer Society.

in differentiation. Our remarks, therefore, will be based mainly upon studies of mammalian spermiogenesis carried out during the past few years in the Department of Anatomy at Cornell.[2] Examples will also be taken from research now in progress on the primitive metazoan, *Hydra*.[3] Because of its small size, the simplicity of its organization, and its remarkable capacity for growth and regeneration, this organism lends itself admirably to investigations with the electron microscope on cell differentiation.

Mitochondria

The submicroscopic internal structure of mitochondria was first described in detail by Palade (1952) after study of a considerable number of tissues representing several animal species. The mitochondria appeared to be bounded by two membranes, of which the outer was smooth and uninterrupted, while the inner was thrown up into narrow folds or *cristae* that projected into a homogeneous matrix in the interior of the organelle. The widespread occurrence of this characteristic structure has been amply confirmed, but as mitochondria of a broader range of vertebrate tissues have been studied, more variation has been found in the abundance and disposition of their internal membranes than was at first envisioned. In some tissues the mitochondria are almost devoid of cristae; in others, they are short, irregularly spaced, and inconstant in their orientation (Plate XIX–1); in still others, numerous long cristae extend nearly across the mitochondrion so that it appears to be traversed by closely spaced, parallel septa (Plate XIX–2). In many protozoa and in a few vertebrate cell types, the internal membranes of the mitochondria are in the form of villi rather than folds and therefore appear in sections as tortuous tubules within the mitochondrial matrix. Thus, while the basic structural plan of mitochondria described by Palade is valid for most tissues, there are minor differences in the pattern of internal membranes that are characteristic of particular cell types.

The population of mitochondria within any one tissue is usually quite uniform, and, to date, reports of significant changes in the internal organization of mitochondria in different stages of differ-

[2] The studies on spermiogenesis have been carried out with the collaboration of Mario Burgos and Susumu Ito.

[3] The observations on *Hydra* are from the current investigations in collaboration with David Slautterback.

entiation of the same cell type have been surprisingly few. Studies on thin sections of the seminiferous tubules of several mammalian species have shown that mitochondria of differentiating germ cells do undergo alterations in shape, changes in intracellular distribution, and a reorganization of their internal structure.

The mitochondria of spermatogonia in the rat and guinea pig are few in number, rod-shaped, and randomly distributed in the cytoplasm. They have the usual internal structure, with numerous cristae oriented more or less perpendicular to the surface. In spermatocytes, mitochondria are more numerous, more nearly spherical, and their cristae are more irregular in their orientation. The mitochondria of early spermatids are generally spherical and their internal structure is atypical, in that the cristae no longer project into the lumen but tend to be folded over and flattened against the inner aspect of the limiting membrane (Plate XIX–4, 5). As a result of this change in disposition of the internal membranes and an accompanying decrease in density of the matrix, the spermatid mitochondria in later stages of differentiation appear to have an angular central cavity bounded by the peripherally displaced cristae. This hollow configuration of the mitochondria persists throughout spermiogenesis. In the spermatids of monkey and man, alterations are also noted in the orientation of the cristae, which are long and become folded upon themselves in such a way that they generally lie parallel to the surface. Because, in these forms, the peripheral displacement of the cristae and the decrease in density of the matrix is less marked than in rodents, no conspicuous central cavity is evident. However, the thin membranes present in the interior of the organelle do have an irregularly concentric arrangement quite unlike anything encountered in the mitochondria of somatic cells. The alterations in mitochondrial structure in the germ cells of the opossum are the most remarkable we have observed. Early in spermatogenesis in this species, the mitochondria are elongated and have the usual pattern of cristae, but in the course of differentiation they are transformed into large spheres completely filled by layer upon layer of concentric membranes (Plate XIX–3). The development of a concentric organization of mitochondrial membranes has also been reported in the spermatids of a snail (Grassé *et al.*, 1956), and in the *nebenkern* of insect germ cells (Beams *et al.*, 1954).

The physiological significance of these alterations in mitochondrial structure during differentiation of the germ cells is obscure. It has been postulated that the cristae serve to increase the area of contact between enzymes, associated with the membranes, and substrates, dispersed in the mitochondrial matrix. It has further been suggested that reaction sequences might be determined by an ordered arrangement of enzymes within the membranes. With these speculations in mind, it is interesting to note that the function as well as the structure of mitochondria may change in the course of spermatid differentiation. Early in spermiogenesis their energy-yielding enzyme systems are probably essential for the synthesis of the acrosome and other specialized structural components of the spermatozoon. Later, their function is presumably concerned with production of energy for locomotion of the sperm. The reorganization observed in the structure of the mitochondria is no doubt related in some way to this change in function.

In addition to the structural alterations in the mitochondria, there are interesting changes in their localization that may also be correlated with a changing role in the economy of the cell. In the early spermatids of the rat, mitochondria congregate in large numbers immediately beneath the cell membrane, leaving very few elsewhere in the cytoplasm. This striking mobilization of mitochondria at the cell surface is a transient event. Later in differentiation, they leave the surface and gather around the base of the flagellum, where they elongate and wrap around the axial bundle of fibrils to form the mitochondrial sheath of the middle piece.

This example of close association between mitochondria and the cell surface is not unique. A somewhat similar relationship is observed in the mammalian nephron, in the Malpighian tubules of insects, and in other epithelia engaged in active transport. In the case of the proximal segment of the nephron, the intimate relation between the two is established by infoldings of the cell membrane that extend inwardly among the mitochondria congregated at the cell base. The mitochondria of the Malpighian tubule migrate out into microvilli that project into the lumen (Beams *et al.*, 1955). In the giant nerve fibers of arthropods these organelles are often located just beneath the axolemma, where, it is believed, they may provide energy for the metabolic exchanges or bioelectric events that occur at the axon–Schwann cell interface (Geren and Schmitt, 1955). The transient mobilization of mitochondria at the surface of the spermatids coincides with the onset of acrosome formation. It may be that in this location they provide energy for rapid importation of metabolites to sustain the synthetic activities of this period of development. Later, when they migrate to the middle piece of the sperm, their proximity to the contractile fibrils of the tail recalls the close juxtaposition of mitochondria and the myofibrils of skeletal muscle. In both situations the mitochondrial enzyme systems doubtless play an essential role in the conversion of chemical energy into mechanical work.

The Endoplasmic Reticulum

Our knowledge of membrane-limited compartments of the cytoplasm other than mitochondria and Golgi apparatus began less than ten years ago, with the first applications of the electron microscope to the study of cells. In this short period the terminology of the cytoplasmic membranes has become so confused and the interpretations of their structure so contradictory, that before presenting our own findings, it will be necessary briefly to review the observational basis for the divergent opinions now held.

Porter and his associates in 1945 observed, in electron micrographs of thin-spread tissue-culture cells, a lacelike network in the cytoplasmic matrix. In later studies this network was defined as a continuous system of anastomosing tubules extending throughout the endoplasm but generally excluded from a thin rim at the periphery of the cell. Irregularly spaced along the course of these membrane-limited canaliculi were numerous small vesicular expansions and occasional broad flat sinuses. In seeking an appropriate name for this system, one that would be descriptive of its netlike arrangement and its continuity throughout the endoplasm, Porter settled on the term *endoplasmic reticulum* (Porter, 1953). Later, when it became possible to study cells in thin sections, the elements of the reticulum appeared as discontinuous profiles of highly variable shape. The tubules and vesicles were round or oval and the large flat sinuses appeared as pairs of parallel membranes separated by a narrow space but continuous with each other at the ends. Components of the latter type, called the *cisternae* of the endoplasmic reticulum (Palade, 1955a), were more abundant in cells *in situ* than they had been in cells *in vitro* and they were often stacked in parallel array, particularly in secretory cells. Also in cells of this kind, small dense granules of uniform size (150 Å) were found adhering to the outer surface of the membranes of the endoplasmic reticulum

(Palade, 1955b). This particulate component of the cytoplasm was subsequently identified as *ribonucleoprotein* (Palade and Siekevitz, 1956). It is now generally accepted that vesicular fragments of this system constitute the microsome fraction of the biochemist (Slautterback, 1953) and that the parallel arrays of cisternae, together with their associated granules, correspond to the basophilic bodies seen with the light microscope in the pancreas and other glands and described by Garnier, in 1899, as the *ergastoplasm*. A number of investigators in this country and abroad have urged the adoption of the term ergastoplasm in place of endoplasmic reticulum. While it is admittedly desirable to retain useful terms from classic cytology, it is undesirable to fit narrowly defined terms to broadly conceived structures. Cytoplasmic basophilia is known to reside in the nucleic acid–rich granules. There are cell types that have the membranous elements of the reticulum well developed but lack the granules and hence are acidophilic. Conversely, there are basophilic cell types in which the two components coexist, but the granules are diffusely distributed in the cytoplasm and not attached to the membranes. The association of the two in glandular cells doubtless has important physiological implications, but to give a separate term, *ergastoplasm*, to the combination of membranes plus RNP granules fails to recognize the fact that this association is merely one particular differentiation of a cytoplasmic organelle that may have a variety of forms and functions in the economy of the cell. If we assign to this structure a name that describes a special case, we may lose sight of the more general significance of the endoplasmic reticulum as a potentially continuous canalicular system permeating the cytoplasm of acidophilic as well as basophilic cell types.

The continuity of the system of cytoplasmic membranes has not been apparent to all investigators, and Sjöstrand, among others, considers the parallel paired membranes in glandular cells to be so different from Porter's original description of the endoplasmic reticulum that there is, in his words, "no justification for homologizing the two" (Sjöstrand, 1956). Instead, he refers to the parallel contours seen in such cells as "double membranes" or "double-edged membranes" and speculates that the space between the two dense lines is occupied, in life, by an organized layer of lipid that is extracted during fixation. Each pair of lines and the hypothetical lipid between are thus thought of as a unit, a three-layered membrane. In putting forward this alternate concept, Sjöstrand denies the existence of a system of channels bounded by membranes and interprets the vesicular and tubular profiles commonly ascribed to such a system as a result of postmortem fragmentation of what he terms *cytomembranes.* Three classes of cytomembranes are distinguished in his terminology. The double contours with surface granules are called α *cytomembranes.* The smooth, double contours that are continuous with the cell surface and that are interpreted in other laboratories as infoldings of the plasma membrane, are called β *cytomembranes.* The membranes of the Golgi complex are termed γ *cytomembranes.* Inasmuch as α *cytomembranes* refer only to the double membranes with adhering granules, this term, like *ergastoplasm,* would seem to be too narrowly and rigidly defined to be conceptually useful. In spite of all the objections that can be marshaled against the term *endoplasmic reticulum,* it is at present the only term sufficiently general in its connotations to cover the dynamic concept of the cytoplasmic membranes that is gradually emerging.

A notable increase in the extent and structural complexity of the endoplasmic reticulum is often seen in the course of differentiation. This is exemplified by the ectodermal cells of the simple metazoon *Hydra,* now being studied in our laboratory by David Slautterback. This interesting organism forms nematocysts that are shot forth to sting and kill its prey. These minute projectiles are produced in clusters of cells called *cnidoblasts.* The primitive cnidoblast has a very sparse endoplasmic reticulum consisting of simple tubular elements. At a slightly later stage of differentiation a more extensive system of tubules and cisternae is found in the neighborhood of the developing nematocyst (Plate XX–1) and, at the height of the synthetic activity of the cnidoblast, the entire cytoplasm is filled with closely packed parallel arrays of cisternae having great numbers of minute ribonucleoprotein granules on their membranes (Plate XX–2 and Plate XXI–1). The appearance of the cytoplasm of these cells is thus very much like that of cells of the pancreas or salivary glands of vertebrates and the elaboration of the nematocyst no doubt involves synthetic activity no less intense than that needed for formation of protein secretory granules.

The sequence of changes in the endoplasmic reticulum of the guinea pig germ cells differs in some respects from that just described. The spermatogonia, like the primitive ectodermal cells of *Hydra,* have only a few tubular and vesicular elements in their cytoplasm. The reticulum undergoes a marked change in form and abundance during the growth and division of the spermatocytes. The predominant form of the reticulum at this stage is that of a network of tubules, but at the periphery extensive cisternae are arranged in layers parallel to each other and to the cell surface (Plate XXI–2). Ribonucleoprotein granules, distributed throughout the cytoplasm, are associated with the membranes only to a limited extent. In early spermatids, a few cisternae are still present but these diminish and disappear as development progresses, leaving a reticulum that consists of a loose plexus of tubular elements. In late spermatids the reticulum becomes more closely meshed and retains a moderately dense amorphous content that is darker than the surrounding cytoplasmic matrix (Plate XXI–3). When the excess cytoplasm is cast off from the elongated spermatid as the residual body of Regaud, the membranes of its endoplasmic reticulum disappear (Plate XXI–4). The ribonucleoprotein particles, however, remain in very great numbers, giving the residual cytoplasm a dense granular character.

Observations on pancreas, salivary glands, plasma cells, and the like have led us to expect the association of ribonucleoprotein granules and parallel arrangement of cisternae in any cell actively engaged in protein synthesis. This expectation was borne out in the case of the cnidoblasts, described earlier. Surprisingly, no such arrangement exists in the spermatids at a time when they are active in the formation of the large acrosome. It should be recalled, however, that the spermatids bear a special relationship to the sustentacular cells of Sertoli and there is reason to believe that these cells may provide not only mechanical support but also products that sustain the differentiation of the spermatids. If this is the case, the role of the endoplasmic reticulum in these cells may be simply to channel such products to the Golgi region where the acrosome is elaborated.

The reteform organization of the endoplasmic reticulum, the apparent canalicular nature of its components, their apparent continuity, and their communication with the perinuclear space and with the flattened vesicular elements of the Golgi complex—all lend credence to the recent suggestions of Porter and Palade that the endoplasmic reticulum may offer preferential pathways for diffusion and may thus constitute an intracellular communication system serving to direct and coordinate biochemical processes within the cell. In mammalian skeletal and cardiac muscle ribonucleoprotein granules are inconspicuous or lacking but the sarcoplasmic reticulum is well developed, forming an elaborate network of canaliculi that surround the individual myofibrils. Here, it has been suggested that the reticulum may have a dual function: the channels may provide pathways for diffusion of metabolites throughout the cell and the membranes themselves may serve to conduct the excitatory impulse (Bennett, 1955; Porter and Palade, 1957). It is likely that the role of the reticulum as an intracellular avenue of communication may prove to be more widespread and may have more general significance than the relation of the membranes and ribonucleoprotein granules to protein synthesis.

In support of our electron-microscope studies on spermatogenesis, Susumu Ito has made valuable parallel observations with the phase contrast microscope on germ cells isolated from guinea pig testis. In living spermatocytes and early spermatids, he has been able to demonstrate phase-dark linear contours that appear to correspond to the cisternae of the endoplasmic reticulum, and has shown that the number of cisternal contours seems to increase as the isolated cells remain under observation. It is strongly suggested that cisternae form under these conditions by coalescence of the tubular and vesicular elements of the reticulum. These findings are in harmony with our current concept in which the reticulum is visualized as an extremely pleomorphic organelle, capable of changing rapidly from one form to another in different physiological conditions. The alterations observed in its form, during successive stages of differentiation within a given cell, doubtless serve to adapt the internal activities of the cell to new functional requirements or changing environmental conditions.

The Golgi Complex

No cell organelle has been subject of more controversy than the Golgi apparatus. Throughout the years of persistent dispute as to whether or not it was an artifact, one of the most cogent arguments for its reality rested upon the reproducible changes in size and form that it displays in different phases of the secretory cycle of glandular cells. At the time the electron microscope was first brought to bear upon this cell component, the majority of cytologists had accepted the existence of the Golgi as a true organelle present in the living cell, but many were of the opinion that the network seen after classic staining methods resulted from coalescence of myelin figures induced by the process of fixation. Of the many observations on the Golgi complex by light microscopists, three merit special consideration in the light of recent findings with the electron microscope. First, a number of investigators noted that the Golgi often appeared to consist of two parts: an outer *osmiophilic* component (Golgi externum) and an inner *osmiophobic* component (Golgi internum) (Hirsch, 1939). Second, reconstruction of the heavily impregnated elements of the Golgi apparatus from serial sections showed that they were not anastomosing canals but flat or curving plates (Pollister, 1939). And finally, the outer chromophilic portion of the Golgi apparatus in certain cell types was reported to show positive birefringence (Monné, 1939), suggesting the presence of oriented structural components. Recent work on the fine structure of the Golgi complex has established that it is made up of lamellar systems of membranes and minute membrane-limited vesicles or vacuoles. The arrays of parallel double membranes seen in electron micrographs no doubt comprised the platelike structures reconstructed by Pollister and the closely spaced parallel arrangement of their component membranes probably accounts for the birefringence observed by Monné. In the conspicuous Golgi zone of developing mammalian germ cells, the parallel membranes tend to be disposed around the periphery, while the small vesicles are located in the interior (Plate XXII–1), an arrangement that corresponds to the distribution of the chromophilic (dictyosome) and chromophobic (idiosome) regions revealed in these same cells by classic Golgi methods. Moreover, it has been shown that the prolonged osmication used in such methods causes a deposition of reduced osmium mainly on the lamellar systems of membranes (Dalton and Felix, 1957).

In light-microscope studies on the elaboration of zymogen granules by the cells of the pancreas, Hirsch and others described the development of vacuoles within an amorphous Golgi ground substance. The contents of the vacuoles gradually increased in density and ultimately were released as secretion granules. Such observations led Hirsch to believe that the Golgi bodies function as sites of segregation and condensation of secretory products formed in other portions of the cell. The results now being obtained with the electron microscope are not inconsistent with this thesis. If we redraw his figure, but substitute for his amorphous Golgi ground substance the parallel arrays of membranes and clusters of minute vesicles that he could not resolve, the resulting diagram (Fig. 27) is in close agreement with published electron micrographs of the Golgi complex in the pancreas and other zymogenic cells (Sjöstrand and Hanzon, 1954).

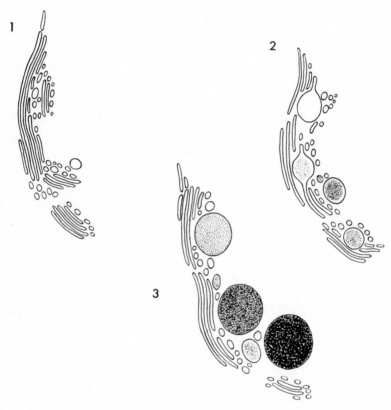

Fig. 27. Diagram of the formation of secretion granules in the Golgi complex. (Redrawn from a familiar diagram of Hirsch, substituting for his amorphous "Golgi ground substance" the parallel double membranes and vesicles revealed by the electron microscope.)

As indicated in the foregoing section of this paper, the endoplasmic reticulum of such cells increases in abundance in the course of their differentiation and comes to be disposed in closely spaced cisternae having numerous small granules on their membranes. There is reason to believe that such lamellar systems with associated nucleic acid–rich granules are sites of active protein synthesis. As a rule, however, no visible product of this synthetic activity accumulates within the "lumen" of the endoplasmic reticulum. Instead, the product, whether it be a secretion granule, an acrosome, or a nematocyst, tends to appear in relation to the Golgi complex as though this were the site of its elaboration. In the early cnidoblast, for example, the endoplasmic reticulum is relatively simple and the Golgi complex is applied to the tip of the developing nematocyst (Plate XX–1). Later, when the nematocyst is rapidly growing, the reticulum is highly developed (Plate XX–2 and Plate XXI–1) but no formed element is visible between its membranes. The Golgi meanwhile remains closely associated with the thin membrane covering the front end of the nematocyst (Plate XX–2). One can speculate that soluble intermediate products are synthesized at the interfaces of the endoplasmic reticulum throughout the cyto-plasm and are channeled through its continuous system of passages to the Golgi region, where they are concentrated into a visible product.

The prominent *idiosome* in the juxtanuclear region of spermato-cytes and spermatids had already been described before the publi-cation of Golgi's classic paper on the reticular apparatus of nerve cells (v. la Valette St. George, 1865). It had also been reported that small granules arose within the idiosome and subsequently coalesced into a single large granule (acrosome) surrounded by a conspicuous vacuole (Moore, 1894; Meves, 1899). After the turn of the century the idiosome gradually came to be identified with the Golgi appara-tus of somatic cells and Bowen (1924), after a detailed study of the formation of the acrosome in the idiosome-Golgi complex, concluded that it was essentially a secretion granule, probably containing en-zymes important in initiating the process of fertilization. In con-sidering the role of the Golgi apparatus in cytodifferentiation, it is pertinent to re-examine Bowen's interpretation of the acrosome, taking advantage of the higher resolution now afforded by the elec-tron microscope. The guinea pig germ cells are particularly favor-able for such a study because of the very large size of their Golgi complex and acrosome.

In primary spermatocytes, the Golgi complex consists of extensive lamellar arrangements of double membranes, which are usually situated at the periphery of the complex (Plate XXII–1), while the interior is occupied by large numbers of minute vesicles from 300 to 500 Å in diameter. Located within the Golgi, or between it and the nucleus, are two cylindrical centrioles often oriented perpendicularly to one another. In secondary spermatocytes, when the Golgi complex becomes active in the formation of acrosomal material, its organization changes considerably (Plate XXII–2). The number of small vesicles increases greatly. These appear to be formed at the expense of the lamellar systems, for the latter become less extensive and clusters of minute vesicles appear around their ends, suggesting that the vesicles are budded off from the edges of the double-membraned lamellae. Associated with these changes, numerous *proacrosomal granules* make their appearance in the interior of the Golgi complex. These vary in density and in size. They are bounded by a membrane and seem to be formed by accumulation of a dense substance within some of the minute vesicles composing the inner zone of the Golgi complex. The contents of the vesicles vary widely in density. Some appear empty; at the opposite extreme, others seem to be solid granules of dense amorphous material enclosed by a membrane. Larger proacrosomal granules seem to arise by coalescence of the smaller granular elements. Although the fate of the proacrosomal granules during the second maturation division has not been followed, it is assumed that they are distributed more or less equally to the daughter cells in golgiokinesis, and that the few large granules found in the Golgi of spermatids have arisen by fusion of proacrosomal granules already present in the late spermatocytes. The process of coalescence continues until there is a single acrosomal granule contained within a sizable vacuole.

The nature of the acrosomal vacuole and its fate in the course of spermiogenesis has been a matter of dispute. In recent studies with the light microscope on testis fixed with Orth's fluid and stained with the periodic-acid Schiff reaction, Leblond and Clermont (1952) did not find a clear area around the granule but instead a less intensely stained, outer zone that they believed to be a part of the acrosomal granule. These able investigators concluded that the vacuole described by earlier workers was an artifact attributable to extraction of the outer zone of the granule by the osmium-containing fixatives employed, and they therefore chose to consider the acrosomal system as a single structure, the *acrosomal granule,* possessing outer and inner zones that differ slightly in their staining reactions. They object to the connotation of emptiness in the term *vacuole* and favor its abandonment. However, it has been found by S. Ito in our laboratory that when isolated living spermatids are examined with phase contrast, a dense spherical acrosomal granule can be

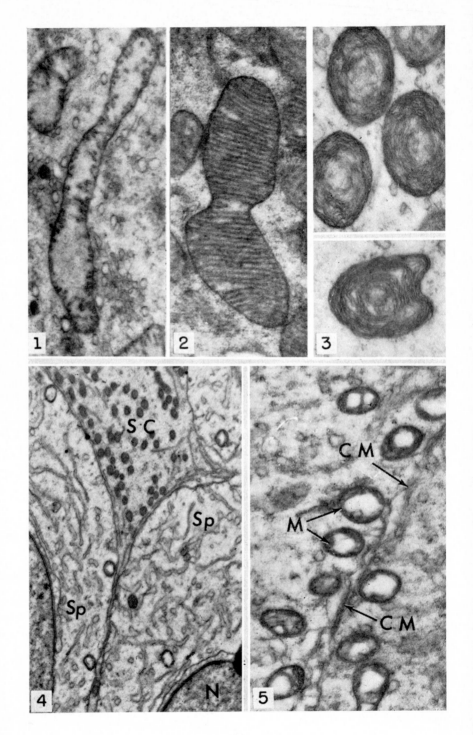

Plate XIX. (1) Mitochondria vary considerably in the complexity of their internal structure. The mitochondrion shown here has villi instead of cristae and these project a very short distance into a matrix of low density. Interstitial cell of the testis (opossum). (2) Mitochondrion with very numerous parallel cristae extending nearly across the organelle. Brown adipose tissue of newborn rat. (*Courtesy,* L. M. Napolitano.) (3) Large spherical mitochondria that show a multilayered concentric arrangement of internal membranes. Late spermatid of opossum. (4) Portions of a Sertoli cell (SC) and two adjacent spermatids (Sp). In the latter, the displacement of the cristae to the periphery results in hollow-appearing mitochondria. Compare these with mitochondria of the neighboring Sertoli cell. Guinea pig testis. (5) Typical spermatid mitochondria (M) with cristae at the periphery more or less parallel with the limiting membrane, leaving a central cavity. Note that the mitochondria are lined up immediately beneath the cell surface (CM). Rat testis.

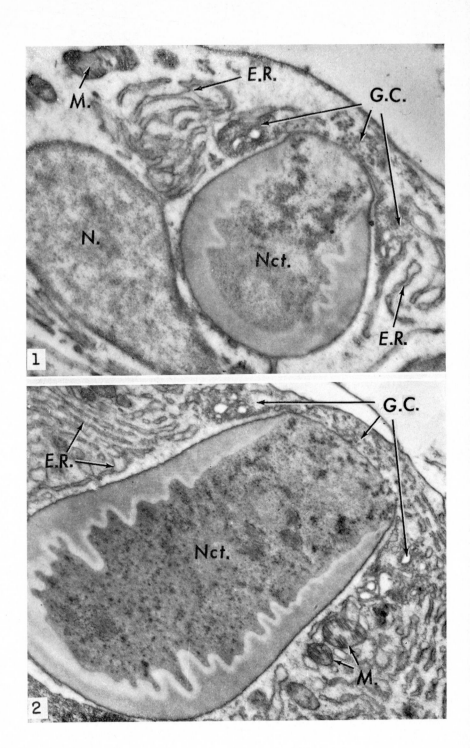

Plate XX. (1) Early cnidoblast in the ectoderm of *Hydra oligactis*. The endoplasmic reticulum (E.R.) is sparse. The Golgi complex (G.C.) is capped over the tip of the developing nematocyst (Nct.). N = nucleus; M = mitochondrion. (*Courtesy*, D. B. Slautterback.) (2) Cnidoblast in a more advanced stage. The endoplasmic reticulum (E.R.) is better developed. The Golgi complex (G.C.) continues to be closely applied to the thin-walled end of the nematocyst (Nct.). The Golgi vesicles appear to coalesce with the limiting membrane of the nematocyst, thus contributing to its contents. (*Courtesy*, D. B. Slautterback.)

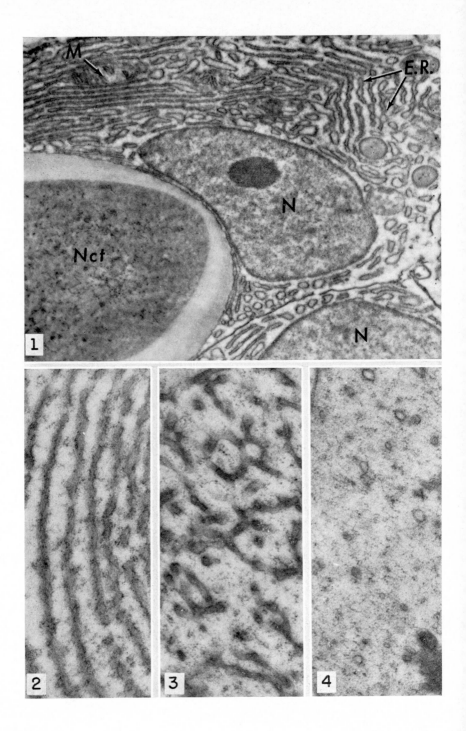

Plate XXI. (1) A cnidoblast of *Hydra* at the height of its synthetic activity. Observe the extraordinary development of the endoplasmic reticulum (E.R.) and compare with the earlier condition, Plate XX–1. Nct = nematocyst; N = nucleus; M = mitochondrion. (2) Parallel cisternae of the endoplasmic reticulum in guinea pig spermatocyte. (3) Endoplasmic reticulum in an advanced spermatid is a network of branching and anastomosing tubules. The ribonucleoprotein granules are not associated with the membranes. (4) The reticulum has largely disappeared in the residual cytoplasm of the late spermatid. The ribonucleoprotein granules persist.

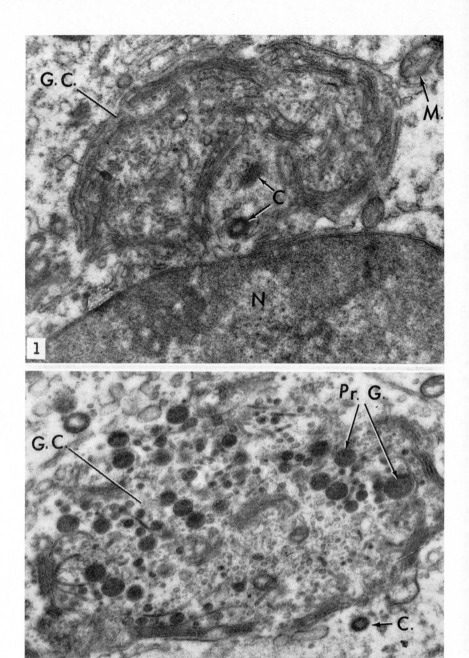

Plate XXII. (1) Relatively inactive Golgi complex (G.C.) of a guinea pig spermatocyte. Lamellar systems of closely spaced membranes are arranged around the periphery in the same distribution as the "dictyosomes" of classic cytology. The remainder of the complex consists of minute vesicles. Two centrioles (C) are situated within the Golgi, one cut transversely and the other tangentially. N = nucleus; M = mitochondrion. (2) Active Golgi complex (G.C.) of late spermatocyte. The lamellar systems of membranes are less conspicuous than before. Great numbers of minute vesicles appear to arise from their edges. Some of the vesicles acquire a dense content and coalesce to form dense proacrosomal granules (Pr.G.). C = centriole.

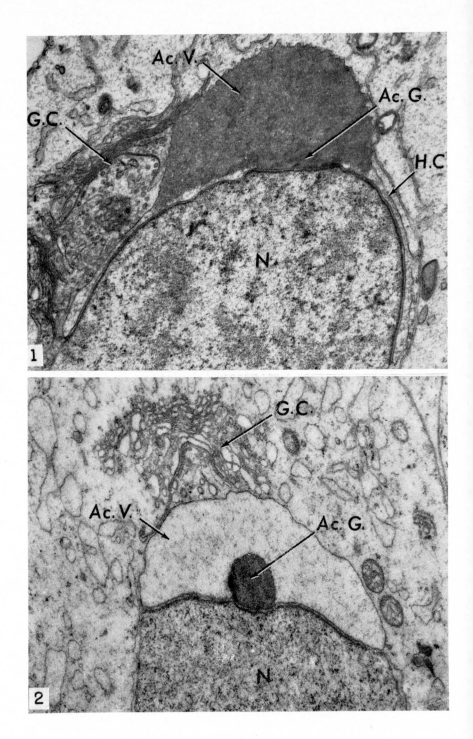

Plate XXIII. (1) Spermatid of a guinea pig in which the acrosomal vacuole (Ac.V.) or vesicle has accumulated a dense material that obscures the outline of the original acrosomal granule. The extension of the limiting membrane of the vacuole down over the nucleus forms the head-cap (H.C.). The Golgi complex (G.C.) has begun to move back toward the caudal end of the spermatid. Ac.G. = acrosome granule; N = nucleus. (2) Spermatid of a cat clearly showing the acrosomal granule (Ac.G.) surrounded by the acrosomal vesicle (Ac.V.). The fluid contents of the vesicle in this species forms only a very light flocculent precipitate upon fixation. The Golgi complex (G.C.) at this stage is still closely associated with the visicle. N = nucleus. (Courtesy, M. H. Burgos.)

seen to float about freely within a sharply defined clear area. This observation on living early spermatids reinforces our belief that at this stage in the evolution of the acrosome there is a *solid* phase (the acrosomal granule) surrounded by a *liquid* phase (the content of the acrosomal vesicle). We continue therefore to use the classic terminology, believing that either *vesicle* or *vacuole* are appropriate terms for a spherical, membrane-limited cytoplasmic structure having a fluid content.

The histochemical observations of Leblond and Clermont demonstrate the presence of a carbohydrate-containing substance dispersed in the fluid content of the acrosomal vesicle. This substance is not entirely lost after osmium fixation but forms a fine-grained flocculent precipitate within the vesicle (Plate XXIII–1, 2), which is most abundant in the guinea pig spermatid. As differentiation proceeds in this species, the acrosomal granule becomes fixed to the limiting membrane of the vesicle in the area where the latter adheres to the nuclear membrane. Thereafter, a progressive increase in the concentration of solids seems to occur within the vesicle, as indicated by phase contrast, which reveals an increasing refractive index of its contents; and by electron micrographs, which show that the vesicle of more advanced spermatids is filled with an amorphous material whose density is only slightly less than that of the granule (Plate XXIII–1). Further inspissation of the material within the vesicle appears to convert it into a homogeneous acrosomal mass indistinguishable from the original granule in refractive index, electron opacity, fine structure, and staining reactions. In other species that develop a smaller acrosome than the guinea pig, the acrosomal granule undergoes relatively little increase in size after it becomes adherent to the nucleus. In osmium-fixed spermatids of these animals no more than a very light precipitate is found in the acrosomal vesicle (Plate XXIII–2). Its fluid contents appear to be absorbed in later development and the limiting membrane of the vesicle then extends down over the sides of the nucleus as the head-cap of the future spermatozoon (Burgos and Fawcett, 1955).

Bowen's conclusion that the acrosome develops within the Golgi complex in much the same manner as a secretion granule is borne out by electron-microscope studies. In the process of acrosome formation in the guinea pig, one sees a particularly striking example of accumulation and concentration of material within a Golgi vacuole. Whether the acrosomal material is synthesized in the Golgi or merely segregated there in visible form cannot be stated.

Coordination of Differentiation

I cannot leave the subject of fine structural changes in differentiation without mentioning recent electron-microscope observations that reveal intercellular connections that may play an important role in coordinating differentiation. In mammalian spermatogenesis the meiotic division of the primary spermatocytes gives rise to pairs of secondary spermatocytes that in turn divide to form groups of spermatids, which then differentiate into the spermatozoa. It is a common observation that within any one group of spermatids, all of the cells are in exactly the same stage of differentiation. The precise synchrony of their development has been attributed to the fact that the spermatids in such a group arise from spermatocytes at about the same time and are enclosed within the same Sertoli cell. When a process of differentiation is initiated at the same moment in several cells sharing the same environment, it would be reasonable to expect the developmental events to proceed on the same time schedule. Study of germinal epithelium with the electron microscope has disclosed another possible explanation. Pairs of spermatocytes are found united by a narrow intercellular bridge and, in the clusters of spermatids resulting from their division, the cells are likewise joined to one another by cytoplasmic bridges (Fig. 28A). These are a micron or more in diameter and are enclosed by a dense membrane that is continuous with the limiting membrane of the conjoined cells (Fawcett and Burgos, 1956). Such intercellular bridges are not to be confused with the slender strands connecting cells in telophase of mitotic division. The latter are believed to result from a transient arrest of cytokinesis when the cleavage furrow encounters remnants of the spindle apparatus. The bridges described here are not so transient but persist until a late stage of spermatid differentiation (Fig. 28B). Although the bridges appear to result from incomplete cytokinesis in the first and second maturation division, they contain no spindle remnants and obviously constitute open communications between cells. Tubular elements of the endoplasmic reticulum very often extend through the bridge, connecting the reticulum of one cell with that of a neighboring cell, and mitochon-

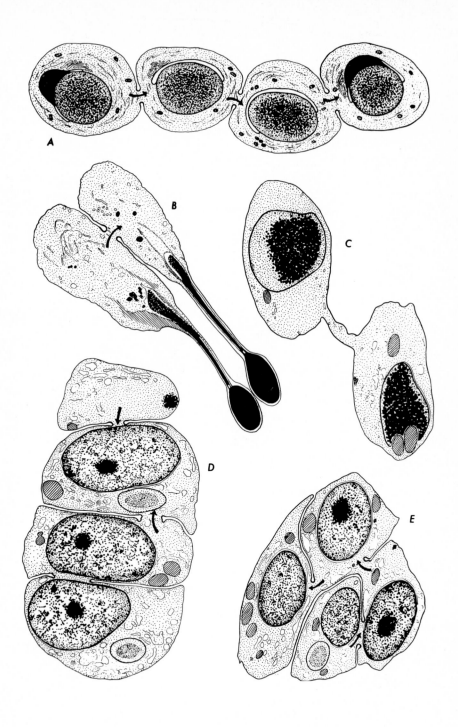

Fig. 28. Several examples of differentiating cells joined by intercellular bridges drawn from electron micrographs. Protoplasmic continuity is believed to be the morphological basis for the synchrony of differentiation exhibited by the cells in such groups. (A) Four conjoined guinea pig spermatids. The two central cells are in the same stage as the two end cells, but their acrosome is out of the plane of this section. (B) Two late spermatids of guinea pig, still joined by a bridge. Presumably two or more others were connected to these in other planes of section. (C) A pair of spermatids from the testis of *Hydra oligactis*. (D) and (E) Batteries of early cnidoblasts from the ectoderm of *Hydra*. As their differentiation proceeds their nematocysts will remain in precisely the same stage of development.

dria may be found lodged between cells. This protoplasmic continuity may be responsible for the coordination of differentiation
within the same group of spermatocytes or spermatids. Intercellular
bridges of this kind have been found connecting the spermatids of
Hydra (Fig. 28C) and those of the rat, guinea pig, rabbit, cat, monkey, and man. It is likely that they will prove to be of general occurrence in developing germ cells throughout the animal kingdom.
They are not confined to the germ cells, however, for in the ectoderm
of *Hydra* it is found that the several cells forming clusters of cnidoblasts are connected by similar bridges (Fig. 28, D and E). The
nematocysts within these conjoined cells are in exactly the same stage
of differentiation. Further study will doubtless disclose additional
examples of this kind and it would not be surprising if intercellular
bridges were found in many places in nature where groups of similar
cells exhibit precisely synchronous differentiation.

References

BEAMS, H. W., and T. N. TAHMISIAN. 1954. The structure of mitochondria in the
    male germ cells of *Helix* as revealed by the electron microscope. *Exper. Cell.
    Res. 6:* 87–93.
BEAMS, H. W., T. N. TAHMISIAN, and R. L. DEVINE. 1955. Electron microscope
    studies on the malpighian tubules of *Melanoplus differentialis*. *Anat. Rec. 121:*
    425.
BENNETT, H. S. 1955. Modern concepts of structure of striated muscle. *Am. J. Phys.
    Med. 34:* 46.
BOWEN, R. H. 1924. On a possible relation between the Golgi apparatus and secre-
    tory products. *Am. J. Anat. 33:* 197.
BURGOS, M. H., and D. W. FAWCETT. 1955. Studies on the fine structure of the
    mammalian testis. I. Differentiation of the spermatids in the cat. *J. Biophys. &
    Biochem. Cytol. 1:* 287–299.
DALTON, A. J., and M. FELIX. 1957. Electron microscopy of mitochondria and the
    Golgi complex. *Symp. Soc. Exp. Biol. 10:* 149–159.
FAWCETT, D. W., and M. H. BURGOS. 1956. Observations on the cytomorphosis of
    the germinal and interstitial cells of the human testis. *Ciba Foundation Coll.,
    Aging 2:* 86–96.
GEREN, B. B., and F. O. SCHMITT. 1955. Electron microscope studies of the Schwann
    cell and its constituents with particular reference to their relation to the axon. Pp.
    251–260. Symposium on Fine Structure of Cells, VIII Congress of Cell Biology,
    Leiden. P. Noordhoff, N. V., Groningen.
GRASSÉ, P., N. CARASSO, and P. FAVARD. 1956. Les ultrastructures cellulaires au
    cours de la spermiogenèse de l'escargot. *Ann. des Sci. Naturelles Zool. 18:* 339–
    380.
HIRSCH, G. C. 1939. Form und Stoffwechsel der Golgikorpen. *Protoplasma Mono-
    graphs.* Berlin.
LEBLOND, C. P., and Y. CLERMONT. 1952. Spermiogenesis of rat, mouse, hamster,
    and guinea pig as revealed by the "periodic acid-fuchsin sulfurous acid" technique.
    Am. J. Anat. 90: 167–215.
MEVES, F. 1899. Uber Struktur und Histogenese der Samenfäden des Meersch-
    weinchens. *Arch. f. mikr. Anat. 54:* 329–402.

MONNÉ, L. 1939. Polarisationsoptische Untersuchungen über den Golgi-Apparat und die Mitochondrien männlicher Geschlechtszellen einiger Pulmonaten-Arten. *Protoplasma 32:* 184–192.

MOORE, J. E. S. 1894. Some points in the spermatogenesis of mammalia. *Internat. Monatschr. f. Anat. u. Physiol. 11:* 129–166.

PALADE, G. E. 1952. The fine structure of mitochondria. *Anat. Rec. 114:* 427–453.

PALADE, G. E. 1955a. A small particulate component of the cytoplasm. *J. Biophys. & Biochem. Cytol. 1:* 59–68.

PALADE, G. E. 1955b. Studies on the endoplasmic reticulum. II. Simple dispositions in cells in situ. *J. Biophys. & Biochem. Cytol. 1:* 567–582.

PALADE, G. E., and P. SIEKEVITZ. 1956. Liver microsomes. An integrated morphological and biochemical study. *J. Biophys. & Biochem. Cytol. 2:* 171–198.

POLLISTER, A. W. 1939. Structure of the Golgi apparatus in the tissues of amphibia. *Quart. J. Micr. Sci. 81:* 235–271.

PORTER, K. R. 1953. Observations on a submicroscopic basophilic component of the cytoplasm. *J. Exp. Med. 97:* 727–750.

PORTER, K. R., A. CLAUDE, and E. F. FULLAM. 1945. A study of tissue culture cells by electron microscopy. *J. Exp. Med. 81:* 233–246.

PORTER, K. R., and G. E. PALADE. 1957. Studies on the endoplasmic reticulum. III. Its form and distribution in striated muscle cells. *J. Biophys. & Biochem. Cytol. 3:* 269–299.

SJÖSTRAND, F. S. 1956. The ultrastructure of cells as revealed by the electron microscope. *Internat. Rev. Cytol. 5:* 455–529.

SJÖSTRAND, F., and V. HANZON. 1954. Ultrastructure of Golgi apparatus of exocrine cells of mouse pancreas. *Exper. Cell. Res. 7:* 415–429.

SLAUTTERBACK, D. B. 1953. Electron microscope studies of small cytoplasmic particles (microsomes). *Exper. Cell. Res. 5:* 173–186.

V. LA VALETTE ST. GEORGE. 1865. Uber die Genese der Samenkörper. *Arch. f. mikr. Anat. 1:* 403–414.

9

Metabolic Interactions in Cell Structures

ALBERT L. LEHNINGER [1]

With the rapidly increasing interest of biochemists in intracellular localization of enzymes and metabolic activities, particularly as afforded by the differential centrifugation procedure for isolating intracellular structures, it is now possible to visualize the beginning of some understanding of the metabolic interactions existing not only *between* various subcellular structures, but also *within* given subcellular structures. Information of this type is obviously of the greatest significance in understanding the biological control of metabolic activities and ultimately the chemical factors leading to cellular development and differentiation. This paper will consider some aspects of such metabolic and enzymatic interactions that have been approached in recent investigations.

Localization of Metabolic Systems and Functions

Preparatory to consideration of some of the interactions between and within intracellular structures it is necessary to describe briefly the localization of some of the major metabolic systems. No attempt will be made to enumerate the distribution of individual enzymes in different structures, which has been reviewed recently (Schneider and Hogeboom, 1956), but rather to discuss the site of complex, highly organized metabolic systems.

Mitochondria

It is now well recognized from a large number of studies on mitochondria from a wide variety of cell types, that these bodies with-

[1] Department of Physiological Chemistry, The Johns Hopkins School of Medicine, Baltimore, Maryland.

out exception show organized respiratory activity (Green, 1957; Lehninger, 1957). This includes in most cases the ability to catalyze the complex reactions of the Krebs citric acid cycle, with pyruvate or acetate as fuel, and electron transfer from the dehydrogenases via respiratory chain carriers including the cytochromes to molecular oxygen. In virtually every case examined electron transfer along the carrier chain is accompanied by coupled phosphorylation of ADP as a means of conserving energy of oxidation. In mitochondria of many cell types, particularly from liver, kidney, and cardiac muscle, the ability to oxidize long-chain fatty acids to CO_2 and H_2O is also present, via the intermediary formation of coenzyme A esters of fatty acids and acetyl-CoA. Both the Krebs cycle and fatty acid oxidation cycle proceed in a highly organized manner with no obvious bottlenecks and at rates consistent with the conclusion that the mitochondria can account for practically all the respiration of the intact cell and the generation of over 90 per cent of the ATP formed from ADP and phosphate in aerobic cells. These oxidations and phosphorylations proceed in an almost autonomous manner, since isolated mitochondria require only substrate, oxygen, Mg^{++}, phosphate, and ADP for nearly maximal activity.

Considerable evidence has been marshaled recently to show that some elements of active transport may occur in mitochondria. Recent work in a number of laboratories (see Davies, 1954) has revealed that isolated mitochondria are capable of causing a rapid exchange, or even an accumulation against a gradient, of certain electrolytes such as Na^+, K^+, and Mg^{++}, coupled to phosphorylating respiration. In addition, it has also been demonstrated that mitochondria are capable of reversible swelling and contraction by passive absorption and active extrusion of water, possibly through a contractile process in the membrane. This process is similarly linked to respiration and phosphorylation.

Nucleus

In addition to containing the elements of the genetic apparatus, the nuclei of at least some cells appear to be the sole cellular site of certain enzymatic reactions leading to synthesis of nucleotide coenzymes such as diphosphopyridine nucleotide and uridine diphospho-glucose (Schneider and Hogeboom, 1956). It appears conceivable that still other enzymes of this type may be associated with nuclei, where they may be assumed to be in a strategic morphological position to interact enzymatically with the DNA and RNA of the nucleus. The nucleotide coenzymes so formed appear not to be utilized as coenzymes in the nucleus but are required for reactions occurring primarily in the cytoplasm.

Recent work by Allfrey and Mirsky (1957) has demonstrated that nuclei are also capable of a small but significant amount of respiration. These investigators have found that calf thymus nuclei contain a respiratory chain reactive with oxygen to which is coupled some phosphorylation of ADP. Although it has certain similarities with the respiratory chain of mitochondria, in that it is inhibited by cyanide and uncoupled by 2,4-dinitrophenol, its behavior with other uncoupling agents is quite different from that of mitochondria. It is conceivable that this respiratory chain is specifically involved in enzymatic synthesis of nucleoside triphosphates for the use of the nucleus or nucleolus in forming RNA and DNA.

Microsomes

The work of Palade and Siekevitz (1956) has clearly demonstrated that submicroscopic particles obtained in differential centrifugation of tissue homogenates, which have been referred to universally as "microsomes," are in fact fragments of the endoplasmic reticulum, or ergastoplasm, together with the ribonucleoprotein granules associated with this very complex cytoplasmic structure. These fragments are capable of incorporating activated amino acids into microsomal protein under certain circumstances. Concurrently, incorporation of nucleotide precursors into RNA occurs in these fragments (cf. Chantrenne, 1958). In addition some phases of cholesterol biosynthesis appear to take place in microsomes, and they also participate in the synthesis of complex lipids (see also Schneider and Hogeboom, 1956).

It is of special interest that the microsomes as isolated by differential centrifugation have significant respiratory activity, although this is certainly minor in rate compared to that of the mitochondria. Microsomal respiration is, however, qualitatively different from mitochondrial respiration, enzymatically speaking. Microsomes contain TPN- and DPN-linked cytochrome reductases, as well as an extramitochondrial cytochrome, namely cytochrome b_5, which is capable of reacting with molecular oxygen, although its action is not inhibited by cyanide (Chance and Williams, 1954; Strittmatter and Velick, 1956). There is no evidence that coupled phosphorylation is associated with this type of extramitochondrial respiration. As Phillips and Langdon (1956) have pointed out, this microsomal respiratory system may serve as a "valve" for controlling the level of reduced pyridine nucleotides outside of the mitochondria, a function which is rather important for the regulation of the reductive biosynthesis of fatty acid, as will be discussed.

The endoplasmic reticulum is vastly complex morphologically and it appears certain that its role in biosynthetic reactions, particularly of protein and RNA, will be found to have a very high degree of enzymatic organization.

Lysosomes

It has been shown that a number of hydrolytic enzymes are found in the so-called mitochondrial fraction as obtained by differential centrifugation. More recently, work of Duve (1957) and others has indicated that this localization is actually due to the presence of another kind of cytoplasmic body present in this fraction, which they have called *lysosomes*. These bodies are characterized by containing such hydrolytic enzymes as uricase, acid phosphatase, β-glucuronidase, ribonuclease, and deoxyribonuclease, the activities of which may be released in soluble form under hypotonic conditions or by treatment with detergents. A fuller appreciation of the metabolic and morphological significance of these bodies must await further work; biologically controlled liberation of hydrolytic enzymes appears to be an attractive possibility.

Cytoplasmic "Sap," or Soluble Fraction

The enzymes involved in the Embden-Meyerhof cycle of glycolysis are now known to be localized very largely in the soluble fraction, in consonance with the fact that these enzymes are easily extractable from various tissues in soluble form (see Schneider and Hogeboom, 1956). The activity of the glycolytic system in this soluble fraction may be greatly altered by the action of other cytoplasmic structures; such control will be discussed in detail below.

The enzymes involved in the so-called "hexose-monophosphate shunt," which may more accurately be called the *pentose-heptose cycle,* are also found in the soluble portion of the cytoplasm. This complex cycle of reactions (see Korkes, 1956), which involves sugars of three, four, five, six, and seven carbon atoms, appears to serve as a mechanism for generating pentose units for nucleic acid and nucleotide synthesis, and as a major source of reduced TPN, which is necessary for fatty acid synthesis and possibly other reductive syntheses (Glock and McLean, 1956). This pentose cycle is most prominent in cell types capable of rapid biosynthesis, such as liver and malignant cells.

Recent work by Langdon (1957) has demonstrated that fatty acid synthesis occurs at very high rates in the soluble fraction of liver homogenates provided suitable supplementary cofactors are present; he has postulated that the soluble fraction is in fact responsible for the greater portion of fatty acid synthesized in the intact cell, on the basis of considerations which will be outlined.

In addition to the organized systems listed above, the soluble fraction contains enzymes responsible for amino acid activation and amino acid transformations, as well as most of the reactions of the urea-arginine cycle.

Some Interactions Between Cell Structures

The foregoing summary of the major metabolic activities occurring in different cellular structures already implies a wealth of possible enzymatic interactions between intracellular structures. However only two multienzyme systems will be discussed briefly to illustrate some mechanisms by which metabolic control is exerted via interactions *between* cell structures, namely, the glycolytic system, which concerns synthesis of carbohydrate and the fatty acid oxidation-synthesis cycle.

Glycolysis

Although it is commonly believed that the glycolytic reactions occur in the soluble portion of the cytoplasm, actually there is substantial evidence that some of the enzymes necessary for the reversible operation of the Embden-Meyerhof cycle are not soluble but are found in other subcellular structures. The hexokinase reaction:

$$(1) \quad ATP + glucose \rightarrow ADP + glucose\text{-}6\text{-}phosphate$$

which is of course the "feeder" reaction of glycolysis, is not found in the soluble fraction of most mammalian cells but is tightly bound to particulate structures (Crane and Sols, 1953), apparently both the mitochondria and "microsomes."

Two ancillary enzymes concerned with the biosynthesis of free glucose from triose phosphates by reversal of the Embden-Meyerhof sequence are also found to be associated with organized structures. Fructose-1, 6-diphosphatase, which catalyzes the reaction

$$(2) \quad \text{fructose-1, 6-diphosphate} + H_2O \rightarrow \text{fructose-6-phosphate} + P_i$$

although found in the soluble fraction of liver, apparently exists as a zymogen or inactive precursor form, which can be activated in a proteolytic reaction by an enzyme present in the microsomes (Pogell and McGilvery, 1952). The activity of this fructose diphosphatase is a function of diet and is greatly increased by cortisone; thus agencies stimulating gluconeogenesis presumably control the microsomal activating enzyme (Mokrasch et al., 1956).

Similarly, glucose-6-phosphatase, catalyzing (3),

$$(3) \quad \text{glucose-6-phosphate} + H_2O \rightarrow \text{glucose} + P_i$$

found only in the liver and kidney and required for formation of the blood sugar, is bound tightly to microsomes. This enzyme is of special interest, since it appears to be under hormonal control; livers of diabetic animals contain much more of the active enzyme than normal livers, and insulin administration causes the level to decline within hours (Langdon and Weakley, 1955). The location of these "valve" enzymes in the microsomes and the fact that they are controlled by endocrine or nutritional factors may be a fact of significance.

Recent work in a number of laboratories (see Krebs, 1954) has established that glyconeogenesis, namely, the synthesis of glucose or glycogen from lactic acid or other nonsugar precursors, does not proceed by direct reversal of the glycolytic cycle as classically written. The direct phosphorylation of pyruvic acid by ATP cannot normally occur at any significant rate under the conditions of concentration of ATP, ADP, and pyruvate which exist in the tissues in situ, because of unfavorable thermodynamic relationships. Rather it has been found that a complementary cycle exists, which involves the interaction of the mitochondria to generate phosphopyruvate from pyruvate by the following series of reactions:

(4)   pyruvate + CO_2 + TPNH → malate + TPN^+
(5)   malate + DPN^+ → oxalacetate + DPNH + H^+
(6)   oxalacetate + inosine triphosphate (ITP) → phospho-enol pyruvate + IDP

These reactions occur in the mitochondria. This complementary cycle makes it possible to by-pass the thermodynamic "block" in the reversal of the glycolytic cycle (Krebs, 1954). The phosphopyruvate thus formed in the mitochondria may readily be converted to glycogen in the soluble cytoplasm by reversal of the remainder of the Embden-Meyerhof cycle.

Clearly, from these considerations, the glycolytic cycle is not a simple series of reactions occurring in the soluble portion of the cytoplasm; it requires in fact a complex interplay between the soluble fraction and the mitochondria and microsomes.

Integration of Glycolysis and Respiration

One of the most interesting problems in intermediary metabolism is the question of the mechanism of the Pasteur effect, namely, the inhibition of glycolysis by respiration so that the rate of glycolysis is evenly matched with the rate of respiration and no accumulation of lactate or pyruvate occurs. Although a great number of mechanisms for the Pasteur effect have been postulated (no attempt will be made to review or assess these here), one of the more significant appears to involve the concentration of inorganic phosphate and ADP in the soluble cytoplasmic compartment (Johnson, 1941; Lynen, 1941). The triosephosphate dehydrogenase reaction of the glycolytic cycle, the second step of which requires inorganic phosphate and ADP as reaction components, may be severely limited when the concentration is very low, as is obvious from the following equations.

(7)   glyceraldehyde-3-phosphate + P_i + DPN^+ $\rightleftarrows$
$$1,3\text{-diphosphoglycerate} + DPNH + H^+$$
(8)   1,3-diphosphoglycerate + ADP $\rightleftarrows$ ATP + 3-phosphoglycerate

Actually the levels of ADP and P_i in the cell, particularly the former, are reflections of the occurrence of oxidative phosphorylation in the mitochondria. As mitochondria oxidize pyruvate they bring about oxidative phosphorylation of ADP. As it happens, the phosphorylating respiration in mitochondria can occur maximally at much lower concentrations of P_i and ADP than are required for maximum reaction rates of the triose phosphate steps in the glycolytic cycle. Mitochondria thus can compete very successfully with triose-phosphate dehydrogenase for the ADP and phosphate necessary to both reactions. Because of this successful competition the rate of the triose-phosphate dehydrogenase step may be inhibited for lack of ADP and thus account for the "braking" effect that respiration exerts on glycolysis. In essence this mechanism is based on the large difference in the Michaelis constants for P_i and ADP of the pertinent enzyme systems in the two metabolic cycles in two different parts of the cytoplasm. Aisenberg and Potter (1957) have presented some extremely interesting evidence for an alternative mechanism, namely, that an intermediate in oxidative phosphorylation in mitochondria (not ATP) controls the activity of hexokinase, which is the "feeder" reaction for glycolysis.

Still another kind of control over the glycolytic cycle by mitochondria may be afforded by regulation of the level of free Mg^{++} in the cytoplasm. It is well known that several reactions of the glycolytic cycle require relatively high concentrations of free Mg^{++} for maximal activity. The Mg^{++} in the cell, however, is present both in the free form and in the form of coordination complexes with various phosphorylated compounds, in particular ATP (see Lardy and Parks, 1956). Raaflaub and Leupin have recently pointed out (1956) that this fact provides the basis for control over the rate of glycolytic reactions by mitochondria and possibly the Pasteur effect. Since the Mg^{++}-ATP complex has a much lower dissociation constant than the Mg^{++}-ADP complex, a high ratio of ATP to ADP in the cytoplasm would tend to tie up most of the cellular Mg^{++} as the Mg^{++}-ATP complex. Such a condition could be brought about by a high rate of respiration in the mitochondria, which in turn produces a high rate of oxidative phosphorylation of ADP to ATP. With cellular Mg^{++} tied up as the Mg^{++}-ATP complex in this manner, the ambient concentration of Mg^{++} in the soluble portion of the cytoplasm may be insufficient for maximal rate of those reactions of the glycolytic cycle that do require relatively high concentrations of free Mg^{++}. Thus the ADP-ATP system is a "Mg^{++} buffer" and can control enzymatic reactions as can ambient pH.

There are many other factors undoubtedly involved in the control of glycolysis in the intact cell; among these may be mentioned the concentration of K^+, the presence of natural coupling or inhibiting factors, and the possibility of association of soluble glycolytic enzymes with cytoplasmic structures in such a subtle manner that the association is broken by the usual homogenization procedure. In any case, the few examples outlined will suffice to demonstrate the point that a very complex degree of interplay may occur between the glycolytic enzymes in the cytoplasm and the various granular cytoplasmic inclusions. The examples that have been described involve types of control occasioned by the properties of the relevant enzymes, such as the Michaelis constants for substrates, metal ions, and cofactors. The basis for this type of enzymatic and metabolic control is thus inherent in the properties of the catalysts involved. In addition to this type of cybernetic interplay, a more subtle and more refined type of control may be superimposed by action of endocrine and organizing agents.

Fatty Acid Oxidation and Synthesis

A rather striking example of the consequences of the cytoplasmic localization of enzymes is afforded by the case of fatty acid oxidation and synthesis. As soon as it was demonstrated that isolated mitochondria can oxidize fatty acids to completion a number of attempts were made to demonstrate that these bodies were also the site of fatty acid synthesis, the underlying argument being that the set of oxidizing enzymes, if reversible in their action, should be able to catalyze both synthetic and oxidative reactions. However, when isolated mitochondria were tested with radioactive acetate under a variety of circumstances, no significant ability to cause fatty acid synthesis was evident. Furthermore when the various enzymes involved in the fatty acid cycle were successfully isolated from mitochondria, brought to a high state of purity, and demonstrated to be reversible in their action, it was found that synthesis of fatty acids would proceed with such a reconstructed system only under the rather special circumstances involved when an artificial electron donor was supplied to the system. However Langdon has demonstrated (1957) that the mitochondria are not the unique site of the individual enzymes involved in fatty acid oxidation. He has shown that the soluble portion of the cytoplasm contains a rather high concentration of all of the enzymes of the fatty acid cycle responsible for the formation of acetyl CoA from longer chain fatty acids. When preparations of the soluble cytoplasmic fraction were suitably fortified with cofactors, in particular reduced triphosphopyridine nucleotide (TPNH), it was found that the soluble fraction would synthesize fatty acids at a very high rate. Thus the mitochondria readily oxidize fatty acids, but have little or no ability to carry out synthesis, although the pertinent enzymes are present; the soluble cell "sap," which has no significant ability to oxidize fatty acids for lack of a cytochrome system, can synthesize them via the action of a separate but similar set of enzymes. Langdon has suggested that the rationale for this situation lies in the ambient oxidation-reduction atmosphere. In the mitochondria there is an active cytochrome system over which electrons from reduced pyridine nucleotides have a very high tendency to flow, for the reduction of molecular oxygen to water. In other words, mitochondria provide a high degree of oxidizing power and it is only reasonable to expect that fatty acid synthesis, which requires the *input* of electrons from pyridine nucleotides, cannot occur. On the other hand, in the soluble portion of the cytoplasm there is no direct contact with the electron transport chain and the reduced pyridine nucleotides are not readily oxidized by intact mitochondria, presumably because of permeability barriers. In the soluble portion of the cytoplasm DPNH is continuously formed during glycolysis and TPNH is continuously formed by action of the pentose cycle. The ratio of reduced to oxidized pyridine nucleotides is thus probably high, whereas it is relatively lower in the mitochondria. This situation presents the opportunity for a very subtle interplay between mitochondria and the soluble portion of the cytoplasm with respect to the balance between fatty acid oxidation and fatty acid synthesis, in which these two processes may occur independently. Compartmentalization of fatty acid oxidation and synthesis is most certainly involved *in vivo;* it may be recalled that in the diabetic animal the ability to oxidize fatty acids is not impaired. On the other hand, it is now well known that one of the striking metabolic defects in the diabetic is the inability to synthesize fatty acids. It is therefore necessary that the hormonal control over oxidation and synthesis be effected in such a manner as to permit synthesis to be controlled without affecting oxidation, a situation produced most readily by the kind of enzymatic and metabolic compartmentalization suggested by Langdon's experiments.

It may be pointed out that Shaw and Stadie (1957) have visualized a similar type of "compartmentalization" of the glycolytic

enzyme system in the rat diaphragm. They have shown that glycogen synthesis from glucose may be stimulated by insulin, without
affecting breakdown of glucose to lactic acid. They explain this
situation by postulating "compartmentalization" of glycolytic enzymes, one set of which is presumed to be involved in the breakdown of glucose to lactic acid, and the other set is believed to be
"compartmentalized" elsewhere in the cell and to function mainly
in the synthesis of glycogen from smaller precursors; the latter is
under more direct hormonal control.

These examples suffice to illustrate the kind of organizational
complexity that will face the student of intermediary metabolism
and cytological organization of metabolism in the future. Although
up to the present the tendency has been to ascribe one or another
metabolic function wholly to one or another cytoplasmic structure,
it is perfectly obvious from the considerations outlined above that
we must expect a more subtle type of enzyme localization and distribution in the intracellular structures.

Intramitochondrial Integration of Respiration, Phosphorylation, and Active Transport

The preceding discussion has considered some mechanisms of
metabolic interplay *between* cytoplasmic structures; in the following, some complex metabolic integration found *within* one cytoplasmic structure, the mitochondrion, will be discussed. It is first
necessary to recall some of the main morphologic and metabolic
features of the mitochondrion. It is now well known from electron-
microscope studies of ultrathin sections of tissues of many different
cell types that the mitochondrion has an outer membrane and an
inner membrane. The latter invaginates into the body of the mitochondrion to form the cristae mitochondriales, although it has also
been suggested that the membranes making up the cristae have no
morphological connection with the outer double membrane. The
membranes apparently consist of monolayers of lipoprotein molecules, which are highly organized enzymatically, as will be shown.
The interior of the mitochondrion contains a soluble matrix, less
dense to the electron beam than the membranes. This is known to
contain soluble proteins and enzymes, a variety of nucleotides and
electrolytes, as well as water. This basic pattern of structure is seen

in mitochondria from many different cell types, both animal and plant.

Intramitochondrial Localization of Enzymatic Activities

It is now possible to say with considerable confidence that the carrier enzymes involved in electron transport between reduced pyridine nucleotide and molecular oxygen, which include a flavo-protein and the cytochromes, are located exclusively in the membranes and/or cristae (Watson and Siekevitz, 1956) in a highly organized lipoprotein structure, which has been extremely refractory to attempts to separate either chemically or mechanically, the individual enzyme proteins making up the respiratory and phosphorylating "assemblies." Perhaps as much as 20 per cent of the mitochondrial membrane may consist of enzyme molecules making up the respiratory chains; it may be calculated that a single mitochondrion contains thousands of these complete respiratory-phosphorylating assemblies. On the other hand, it appears probable that most of the enzymes concerned with the substrate interactions of the tricarboxylic acid cycle are present in the soluble matrix of the mitochondrion, since most of these may be quite easily extracted following disruption of the mitochondria. Similarly the enzymes involved in fatty acid oxidation are also easily extractable in soluble form from mitochondria and hence are probably found in the inner soluble matrix. Not all dehydrogenases, however, are found in this matrix; both succinic dehydrogenase and D-β-hydroxybutyric dehydrogenase are apparently attached rather firmly to the respiratory enzyme assemblies found in the mitochondrial membrane (Gamble and Lehninger, 1956).

The location of the respiratory chain enzymes in the mitochondrial membranes, together with the extremely important enzymatic mechanisms required for conversion of the energy of oxidation into phosphate bond energy via oxidative phosphorylation can therefore be expected to have a very great bearing on those functions of mitochondria in which passage of substances through the mitochondrial membrane is involved, namely, the active movement of water and electrolytes across this membrane. As will be seen in the following discussion, the enzymatic activity of the respiratory assemblies that are embedded in the mitochondrial membranes can control passage of substances in a very dramatic fashion.

Active Movement of Water

Work in a number of laboratories has led to the following picture of active movement of water accompanying oxidation and phosphorylation (see Price *et al.*, 1956). When respiration or phosphorylation is inhibited, mitochondria of the liver and kidney, for example, will swell in a passive manner with an increase in water content, which can be measured either directly or indirectly by following the optical density or light scattering of suspensions of mitochondria. A number of factors can influence the rate of this spontaneous swelling of mitochondria, which is apparently passive in nature. For instance, inorganic phosphate, calcium ions, and certain sulfhydryl reagents cause a great increase in the rate of this spontaneous swelling (see Raaflaub, 1953; Tapley, 1956). On the other hand, a number of substances prevent this swelling phase by inhibiting the rate of water entry. Among these are ATP and other nucleoside di- and triphosphates, magnesium or manganese ions, and certain metal binding agents such as ethylenediamine tetraacetate, which apparently act through binding calcium ions. The effect of a variety of agents known to uncouple oxidative phosphorylation has been tested. It has been found rather unexpectedly that such agents as dinitrophenol, pentachlorophenol, and Dicumarol all inhibit the spontaneous swelling. On the other hand, thyroxine and its analogs greatly increase the rate of the spontaneous swelling reaction, a finding that has led to a new approach to the mode of action of this hormone, which will be outlined (Lehninger, 1956).

In addition to the passive phase of mitochondria swelling noted above, there is also a less well-understood active phase in which water is extruded during respiration and phosphorylation (Price *et al.*, 1956). However the active water extrusion has been found to be a delicately poised function that is rather erratic when studied *in vitro*. The mitochondria must be isolated under the best circumstances with a high degree of morphological intactness; demonstration of the active extrusion phase *in vitro* requires conditions in which the passive swelling phase has not gone beyond the "point of no return." Despite these difficulties, sufficient observations are now at hand to suggest that the mitochondria are normally in a dynamic steady state in which passive diffusion of water into the mitochondrial structure is counterbalanced by respiration-linked extrusion of water and that interference with these processes can lead to "cloudy swelling" of the mitochondria. Under very special circumstances it has been found that ATP alone, without the requirement of respiration and oxidative phosphorylation, will cause some active contraction of mitochondria. However not all workers have been able to observe such effects.

Because the amounts of water exchanged under conditions of active extrusion are relatively large on a molar basis compared to the amount of ATP that might be split during such a process, it has been suggested (Raaflaub, 1953; Price et al., 1956) that the mitochondrial membrane has contractile properties similar to those of the actomyosin system of myofibrils; the passive and active phases of swelling and active extrusion of water seem to be conditioned by interaction with phosphorylated compounds such as ATP. Because of the close similarities between mitochondrial ATPases and the ATPase activity of myosin or actomyosin systems, this idea may prove to be an extremely fruitful one in approaching the enzymatic organization of the mitochondrial membrane.

Active Movement of Electrolytes

Work in a number of laboratories has revealed that isolated mitochondria from liver or kidney, while actively respiring and phosphorylating, are capable of increasing the rate of exchange of certain electrolytes, particularly potassium, or of causing actual accumulation of these ions against a gradient. These active transport mechanisms are clearly coupled to oxidation and phosphorylation, since inhibition of respiration or uncoupling of phosphorylation inhibits these processes. Active accumulation of electrolytes and other substances must involve two considerations: (1) the actual permeability of the mitochondrial membrane, and (2) the activity of some system within the mitochondrial body for causing selective accumulation of the substance in question. The mechanism of such accumulation reactions, and how they are geared to active respiration and phosphorylation, are still quite obscure. Recently Gamble (1957) in our laboratory has provided an interesting experimental lead to the study of this accumulation process on a molecular basis. Using fragments of the mitochondrial membrane, which contain phosphorylating respiratory assemblies, he has found that while respiration and phosphorylation occur in such particles, there is a concomitant active binding of radioactive K^+ to the fragments, which can be detected by centrifuging them out of the actively respiring system and counting the bound K^{42}. The active binding so observed can be inhibited by respiratory inhibitors such as cyanide or 2,4-dinitrophenol. These findings suggest that the membranes, possibly the inner one forming the cristae, are active in the ion transport process.

Action of Thyroxine on Mitochondria

A number of effects of thyroxine on the activity of certain enzymes have been described in the literature and have been used to furnish possible explanations for the physiological action of this hormone. The most striking of these observations is the fact that under certain experimental conditions the *in vitro* addition of thyroxine and its analogs can cause the uncoupling of oxidative phosphorylation in intact mitochondria. Because of the pronounced metabolic effects of thyroxine with respect to the basal metabolic rate, this effect has appeared to many investigators to furnish the most likely physiological explanation for the action of this hormone.

We have been led to a somewhat different picture (see Lehninger, 1956) of the mode of action of thyroxine through studies of the effect of uncoupling agents on the phosphorylating mitochondrial membrane fragments previously described. Although dinitrophenol uncouples phosphorylation in both intact mitochondria and in the mitochondrial membrane preparations, thyroxine has a somewhat different action. It is capable of uncoupling phosphorylation in more or less intact mitochondria, particularly after hypotonic treatment of the latter, but is completely inert when it is tested against phosphorylation as it occurs in suspensions of fragments of the mitochondrial membrane (Cooper and Lehninger, 1956; Tapley and Cooper, 1956). This has led us to the view that thyroxine has some effect on mitochondrial structure that may cause uncoupling of phosphorylation in intact mitochondria but that is not occasioned by direct interaction of thyroxine with the enzymes required for coupling phosphorylation to oxidation. The effect of thyroxine on some of the membrane reactions of mitochondria were then examined, and it was soon found by Tapley in our laboratory (1956) that thyroxine causes a pronounced swelling of mitochondria in the absence of phosphorylation or oxidation; in other words, thyroxine greatly increases the rate of the passive phase in which water is taken up by the mitochondria. Furthermore these swelling effects of thyroxine on the mitochondria are produced by concentrations of thyroxine considerably below those required to produce uncoupling of phosphorylation. Dinitrophenol, on the other hand, has a diametrically opposed effect; it prevents the spontaneous swelling of mitochondria and in fact can counteract the swelling effect of thyroxine.

This action of thyroxine has been more thoroughly studied recently (Lehninger *et al.*, 1958, 1959), and some characteristics of the reaction may be briefly outlined. The swelling reaction produced by thyroxine is extremely sensitive to pH, having an optimum at about pH 7.4. There is no detectable swelling at pH 9 or pH 6.5. The swelling reaction induced by thyroxine has also been found to have an extremely high temperature coefficient; whereas most chemical and enzymatic reactions have a temperature coefficient of approximately 2, the swelling reaction has a temperature coefficient as high as 4 or 5. Although the significance of this high temperature coefficient is not entirely known, such a high coefficient is found in many membrane reactions and may indicate that the reaction is a result of the interplay of many factors in which a highly complex structure undergoes a change to a much more random configuration. The swelling action of thyroxine can be demonstrated in concentrations of the order of $10^{-8}M$, which is approximately the concentration of thyroxine occurring in the blood and tissues *in vivo*. This *in vitro* action of thyroxine is thus brought about at physiological concentrations of the hormone; no other *in vitro* effects of thyroxine have been observed to have such a high sensitivity.

The swelling action of thyroxine on mitochondria is integrated with the enzymes of respiration and phosphorylation. Normally the thyroxine-induced swelling is observed under aerobic circumstances and is apparently maximal under these conditions (it must be stated again that *net* respiration or phosphorylation is not taking place in such passive swelling experiments). However, if the mitochondria are brought into anaerobic circumstances, then the swelling action of thyroxine is completely inhibited (Lehninger and Ray, 1957). Under these experimental conditions the respiratory carriers, namely, pyridine nucleotides, flavoprotein, and the various cytochromes, are present in the mitochondrial membrane in the reduced state, whereas under aerobic circumstances these carriers are entirely in the oxidized state. Since the respiratory enzyme assemblies make up almost 20 per cent of the mitochondrial membrane, which in turn has enzymatically controlled contractile properties, it is not surprising that the properties of the membrane should be greatly altered by changes in the oxidation-reduction state of the carriers. These effects on mitochondrial swelling correlate with the action of phosphorylating enzymes (Wadkins and Lehninger, 1957).

As Chance and Williams have shown (Chance and Williams, 1956), the respiratory carriers of intact mitochondria are in a dynamic steady state, even when they are actively respiring. Under such circumstances the carriers at the substrate end of the chain are more fully reduced, whereas the carriers near the oxygen end of the chain are in the more fully oxidized state. Obviously the response of mitochondria in the intact cell to ambient thyroxine must be conditioned by the supplies of substrate, oxygen and ADP and phosphate, to form an extremely complex interplay of metabolic and hormonal controlling factors over the rate of water uptake by mitochondria. The action of thyroxine on the active phase of water extrusion has not yet been fully examined.

These *in vitro* observations on the action of thyroxine on mitochondrial swelling have been substantiated by a study of mitochondria isolated from hypo- and hyperthyroid rat livers compared to normal mitochondria (Tapley, 1956). It was found that mitochondria isolated from hyperthyroid rat liver are very fragile and have a very much higher rate of swelling under standard test conditions than normal mitochondria. Conversely, mitochondria isolated from hypothyroid rat livers are much more resistant to swelling when tested *in vitro*. These observations are of great assistance in approaching the physiological mechanism for the action of thyroxine, which, as can be seen from the preceding account, is not simply an uncoupling of phosphorylation similar to that brought about by dinitrophenol but a very delicate and subtle effect on the condition of the mitochondrial membrane.

These observations on the various metabolic and enzymatic reactions of isolated mitochondria illustrate the high degree of complexity of metabolic interaction *within* a cytoplasmic structure and serve to illustrate the kind of considerations, both structural and enzymatic, which must enter into the analysis of metabolic interactions in other cytoplasmic components.

References

AISENBERG, R., and V. R. POTTER. 1957. Studies on the Pasteur effect. *J. Biol. Chem.* 224: 1099–1127.

ALLFREY, V. G., and A. E. MIRSKY. 1957. The role of deoxyribonucleic acid and other polynucleotides in ATP synthesis by isolated cell nuclei. *Proc. Natl. Acad. Sci.* 43: 589–598.

CHANCE, B., and G. R. WILLIAMS. 1954. Kinetics of cytochrome b_5 in rat liver microsomes. *J. Biol. Chem.* 209: 945–951.

CHANCE, B., and G. R. WILLIAMS. 1956. The respiratory chain and oxidative phosphorylation. *Advances Enzymol.* 17: 65–134.

CHANTRENNE, H. 1958. Newer developments in relation to protein biosynthesis. *Ann. Rev. Biochem.* 27: 35–56.

COOPER, C., and A. L. LEHNINGER. 1956. Oxidative phosphorylation by an enzyme complex from extracts of mitochondria. *J. Biol. Chem.* 219: 489–506.

CRANE, R. K., and A. SOLS. 1953. The association of hexokinase with particulate fractions of brain and other tissue homogenates. *J. Biol. Chem.* 203: 273–292.

DAVIES, R. E. 1954. Relations between active transport and metabolism in some isolated tissues and mitochondria. In *Active Transport,* Symp. Soc. Exper. Biol. 8. Academic Press, Inc., New York. Pp. 453–475.

DUVE, C. 1957. The enzymic heterogeneity of cell fractions isolated by differential centrifugating. In *Mitochondria,* Symp. Soc. Exper. Biol. 10. Academic Press, Inc., New York. Pp. 50–61.

GAMBLE, J. L., JR. 1957. Potassium binding and oxidative phosphorylation in mitochondria and mitochondrial fragments. *J. Biol. Chem.* 228: 955–971.

GAMBLE, J. L., JR., and A. L. LEHNINGER. 1956. Activity of respiratory enzymes and adenosine triphosphatase in fragments of mitochondria. *J. Biol. Chem.* 223: 921–933.

GLOCK, G. E., and P. MCLEAN. 1955. The determination of oxidized and reduced diphosphopyridine nucleotide and triphosphopyridine nucleotide in animal tissues. *Biochem. J.* 61: 381–390.

GREEN, D. E. 1957. Organization in relation to enzymic function. In *Mitochondria,* Symp. Soc. Exper. Biol. 10. Academic Press, Inc., New York. Pp. 30–49.

JOHNSON, M. J. 1941. Role of aerobic phosphorylation in the Pasteur effect. *Science* 94: 200–203.

KORKES, S. 1956. Carbohydrate metabolism. *Ann. Rev. Biochem.* 25: 685–734.

KREBS, H. A. 1954. Considerations concerning the pathways of synthesis in living matter. *Bull. Johns Hopkins Hosp.* 95: 19–33.

LANGDON, R. G. 1957. The biosynthesis of fatty acids in rat liver. *J. Biol. Chem.* 226: 615–630.

LANGDON, R. G., and D. R. WEAKLEY. 1955. The effect of hormonal factors and of diet upon hepatic glucose-6-phosphatase activity. *J. Biol. Chem.* 214: 167–174.

LARDY, H. A., and R. E. PARKS, JR. 1956. Influence of ATP concentration on rates of some phosphorylation reactions. In *Enzymes: Units of Biological Structure and Function.* Academic Press, Inc., New York. Pp. 584–587.

LEHNINGER, A. L. 1956. Physiology of mitochondria. In *Enzymes: Units of Biological Structure and Function.* Academic Press, Inc., New York. Pp. 217–234.

LEHNINGER, A. L. 1957. Relation of oxidation, phosphorylation, and active transport to the structure of mitochondria. In *Molecular Biology.* University of Chicago Press, Chicago. (In press).

LEHNINGER, A. L., and B. L. RAY. 1957. Oxidation-reduction state of rat liver mitochondria and the action of thyroxine. *Biochim. et Biophys. Acta.* (In press).

LEHNINGER, A. L., B. L. RAY, and M. SCHNEIDER. 1959. The swelling of rat liver mitochondria by ATP and its reversal. *J. Biophys. Biochem. Cytol.* (In press).

LEHNINGER, A. L., C. L. WADKINS, C. COOPER, T. M. DEVLIN, and J. L. GAMBLE, JR. 1958. Oxidative phosphorylation. *Science. 128:* 450–456.

LYNEN, F. 1941. Aerobic phosphate requirements of yeast-Pasteur reaction. *Ann. Chem. 546:* 120–133.

MOKRASCH, L. C., W. D. DAVIDSON, and R. W. McGILVERY. 1956. The response to glucogenic stress of fructose-1,6-diphosphatase in rabbit liver. *J. Biol. Chem. 222:* 179–184.

PALADE, G. E., and P. SIEKEVITZ. 1956. Liver chromosomes. An integrated morphological and biochemical study. *J. Biophys. Biochem. Cytol. 2:* 171–187.

PHILLIPS, A. H., and R. G. LANGDON. 1956. The effect of thyroxine and other hormones on hepatic TPN-cytochrome reductase activity. *Biochim. et Biophys. Acta 19:* 380–382.

POGELL, B. M., and R. W. McGILVERY. 1952. The proteolytic activation of fructose-1,6-diphosphatase. *J. Biol. Chem. 197:* 293–302.

PRICE, C. A., A. FONNESU, and R. E. DAVIES. 1956. Movements of water and ions in mitochondria; The prevention of swelling of mitochondria. *Biochem. J. 64:* 756–776.

RAAFLAUB, J. 1953. Die Schwellung isolierter Leberzellmitochondrien und ihre physikalisch-chemische Beeinfluszbarkeit. *Helv. physiol. et pharmacol. acta 11:* 142–157.

RAAFLAUB, J., and I. LEUPIN. 1956. Use of metal-buffers in enzyme reactions, pMg-activity curve of hexokinase from yeast. *Helv. chim. acta 39:* 832–843.

SCHNEIDER, W. C., and G. HOGEBOOM. 1956. Biochemistry of cellular particles. *Ann. Rev. Biochem. 25:* 201–224.

SHAW, W. N., and W. C. STADIE. 1957. Coexistence of insulin-responsive and insulin-non-responsive glycolytic systems in rat diaphragm. *J. Biol. Chem. 227:* 115–134.

STRITTMATTER, P., and S. F. VELICK. 1956. The isolation and properties of microsomal cytochrome. *J. Biol. Chem. 221:* 253–264; A microsomal cytochrome reductase specific for diphosphopyridine nucleotide. *J. Biol. Chem. 221:* 277–286.

TAPLEY, D. F. 1956. The effect of thyroxine and other substances on the swelling of isolated rat liver mitochondria. *J. Biol. Chem. 222:* 325–339.

TAPLEY, D. F., and C. COOPER. 1956. The effect of thyroxine and related compounds on oxidative phosphorylation. *J. Biol. Chem. 222:* 341–349.

WADKINS, C. L., and A. L. LEHNINGER. 1957. Oxidation state of respiratory carriers and the mechanism of oxidative phosphorylation. *J. Am. Chem. Soc. 79:* 1010.

WATSON, M. L., and P. SIEKEVITZ. 1956. Cytochemical studies of mitochondria. *J. Biophys. Biochem. Cytol. 2:* 639–652.

Index